SCIENCE
A CLOSER LOOK

BUILDING SKILLS

Activity Lab Workbook

Macmillan/McGraw-Hill

Contents

Contents

Dear Parent or Guardian,

   Today our science class talked about how to work safely when doing laboratory experiments. It is important that you be informed regarding the school's effort to promote a safe environment for students participating in laboratory activities. Please review the safety rules and this entire Safety Contract with your child. This contract must be signed by both you and your child in order for your child to participate in laboratory activities.

Safety Rules:

1. Listen carefully and follow directions.

2. Perform only those experiments approved by your teacher. If you are not sure about something, ask your teacher.

3. Take great care when handling and moving chemicals and hot materials.

4. Conduct yourself in a responsible manner at all times.

5. Always clean up after you have finished an experiment.

6. Always wash your hands before and after an experiment.

7. Do not eat, drink, or chew gum in the laboratory.

Date: _____

   I have read and reviewed the science safety rules with my child. I consent to my child's participation in science laboratory activities in a classroom environment where these rules are enforced.

   Parent/Guardian signature: _____

   I know that it is important to work safely in science class. I understand the rules and will follow them.

   Student signature: _____

Estimados padres o tutor:

Hoy hemos hablado en nuestra clase de Ciencias sobre cómo mantener la seguridad al realizar experimentos científicos. Es importante que ustedes estén informados del propósito de la escuela de promover un entorno seguro para los estudiantes que participan en las prácticas de laboratorio. Por favor, examinen cuidadosamente con su niño o niña las reglas siguientes y el Acuerdo de Seguridad. El acuerdo debe ser firmado tanto por uno de ustedes como por su niño o niña para que él o ella pueda participar en las actividades de laboratorio.

Reglas de Seguridad:

1. Escucha con atención y sigue las indicaciones.

2. Haz sólo los experimentos aprobados por tu maestro o maestra. Pregúntale a él o a ella si no estás seguro de algo.

3. Ejercita sumo cuidado al manipular y transportar productos químicos y materiales calientes.

4. Compórtate en todo momento de manera responsable.

5. No te olvides de limpiar cuando termines de realizar un experimento.

6. Lávate siempre las manos antes y después de hacer un experimento.

7. No comas, bebas ni mastiques chicle en el laboratorio.

Fecha: _____

He leído y examinado las reglas de seguridad de ciencias con mi niño o niña. Doy mi consentimiento para su participación en las actividades del laboratorio de ciencias en un entorno donde se hagan cumplir estas reglas.

Firma de uno de los padres o tutor: _____

Sé la importancia que tiene trabajar con seguridad en la clase de Ciencias. Comprendo las reglas y me comprometo a seguirlas.

Firma del estudiante: _____

What do you know about animals that live in Madagascar?

Meet two scientists who are curious about the natural world and everything that lives in it. Chris Raxworthy and Paule Razafimahatratra study animals that live in Madagascar. They work at the American Museum of Natural History in New York City and at the University of Antananarivo in Madagascar.

Use the text in your book to help you answer the questions below.

1 How would you look for animals in their natural habitat?

_____

_____

2 What kinds of animals would you see in the forest?

_____

_____

3 What does an animal need to live in the forest?

_____

_____

4 How do scientists find answers to these questions?

_____

_____

Name _____ Date _____

Draw Conclusions

5 What do scientists do?

_____

_____

_____

6 How do scientists test a hypothesis?

_____

_____

Explore More

How do scientists draw conclusions?

_____

_____

_____

Open Inquiry

Think of your own question about why animals live in different places and different forests. Make a plan and carry out an experiment to answer your question.

My question is: _____

_____

How I can test it: _____

_____

My results are: _____

_____

**Be a Scientist
Activity Lab Book**

What do you know about studying animals?

Materials

• reference materials such as an encyclopedia or the Internet

Procedure

1 Explore more about animals by identifying the animals that live in your neighborhood and using reference materials to research facts about them. Then answer these questions.

2 What kinds of animals, besides pets, live in your neighborhood?

_____

3 Where do these animals find food, water, and shelter?

_____

_____

4 Suppose you wanted to learn more about one of these animals. What would you do?

_____

_____

_____

5 Imagine you were going with Chris and Paule to a forest in Madagascar to study animals. What things would you bring?

_____

_____

© Macmillan/McGraw-Hill

Name _____ Date _____

How do living and nonliving things differ?

Purpose

Find out some characteristics of living and nonliving things.

Procedure

1 **Predict** How are all living things alike? How are nonliving things alike?

_____

_____

2 Make a table on a separate piece of paper. Label the columns *Living Things* and *Nonliving Things*.

3 Place 4 pieces of string outside on the ground so that they form a square.

4 **Observe** Look for living things in your square area. List them in your table. Tell how you know they are living. Do the same with nonliving things that you see.

Step **4**

Draw Conclusions

5 **Interpret Data** What characteristics do the living things share? Which do the nonliving things share?

_____

_____

_____

_____

Explore More

Experiment Does the amount of sunlight affect how many living things are in an area? How could you test this?

_____

_____

_____

Open Inquiry

Design additional activities to differentiate between living and nonliving things.

My question is: _____

_____

How I can test it: _____

_____

_____

My results are: _____

_____

Is a shell alive?

Purpose

In this activity, you will look at shells to find out if they are alive.

Procedure

1 **Observe** Use the hand lens to look at at least two different shells.

2 **Communicate** In the space below, draw two of the shells you observed.

3 **Communicate** List the characteristics of the shells you observed.

_____

_____

Draw Conclusions

4 **Infer** Do you think that shells are alive? Why or why not?

_____

_____

© Macmillan/McGraw-Hill

Observe Cells

1 **Observe** Look at a piece of onion. Then observe it using a hand lens. What do you see?

_____

_____

2 **Communicate** On a separate piece of paper, draw how the onion looks when viewed with a hand lens.

3 **Observe** Look at a slide of an onion under a microscope. What do you see? Is there any space between the cells?

_____

_____

4 **Communicate** Draw how the onion looks when viewed with a microscope. Then compare your two drawings.

5 **Infer** How small are cells? What tool do you need to observe cells?

_____

_____

Name _____ Date _____

How are plants alike?

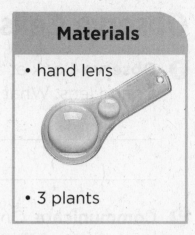

Materials
- hand lens
- 3 plants

Purpose

Find out about some characteristics of plants.

Procedure

1 **Observe** Look carefully at each plant. Which plants have leaves? How do their leaves compare? Describe them.

_____

_____

2 **Infer** Which part of each plant grows underground? How is this part the same on each plant? How is it different?

_____

_____

Step **1**

Step **2**

3 **Observe** Look carefully at each plant again. What other parts does each plant have? Record your observations.

_____

_____

Draw Conclusions

4 Infer What parts do most plants have?

_____

_____

Explore More

Experiment Can different-looking plants survive under the same conditions? How could you find out? Make a plan and try it.

_____

_____

_____

Open Inquiry

Design additional activities to find out what plants need to grow.

My question is: _____

_____

How I can test it: _____

_____

My results are: _____

_____

How are leaves different?

Purpose

In this activity you will observe various types of leaves and compare their characteristics.

Procedure

1 **Observe** Use the hand lens to look at several different kinds of leaves.

2 **Communicate** Record your observations in the chart below.

Type of Leaf	Leaf Shape	Leaf Size	Leaf Color

3 **Infer** How do you think the different leaves help plants?

_____

_____

_____

Draw Conclusions

4 **Communicate** What characteristics do all of the leaves have in common?

_____

_____

© Macmillan/McGraw-Hill

Observe Stems

1. Get a stalk of celery with leaves on it. Carefully cut two inches off of the bottom.

2. Half fill a plastic jar with water. Then add five drops of food coloring to the water. Mix the water with a spoon.

3. **Observe** Place the celery into the jar. Observe the celery stalk a few times throughout the day. What do you notice?

   _____

   _____

4. **Communicate** How has your celery stalk changed? Draw a picture. Write a description.

   _____

   _____

5. **Infer** What do stems do?

   _____

   _____

Structured Inquiry

What do plants need to survive?

- 4 identical plants

- measuring cup and water

- ruler

Form a Hypothesis

Do plants need light to grow? Do they need water? Write a hypothesis. Start with, "If plants do not get light and water, then . . ."

_____

_____

Test Your Hypothesis

Step 2

1. Label four identical plants as shown.

2. **Observe** How do the plants look? How tall are they? Measure them and record your observations in a chart. Use words and pictures.

_____

_____

3. Put the plants labeled *No Light* in a dark place, such as a closet. Put the plants labeled *Light* in a sunny place, such as on a windowsill.

4. **Predict** What do you think will happen to each plant? Record your predictions.

_____

_____

5 **Observe** Look at the plants every other day. Water each plant labeled *Water* with 200 mL of water. Measure how tall the plants grow. Record your observations in your chart using words and pictures.

Step **5**

Draw Conclusions

6 **Interpret Data** Which plant grew the most after two weeks? Which plant looks the healthiest? Use your chart to help you.

_____

_____

7 What do plants need to survive?

_____

_____

Name _____ Date _____

Guided Inquiry

What else do land plants need to survive?

Form a Hypothesis

Do plants need air? Do they need nutrients? Write a hypothesis about one of these.

_____

_____

_____

Test Your Hypothesis

Design an experiment to test your hypothesis. Decide which of the materials below you will use. Write the steps you will follow on a separate piece of paper.

▶ two identical plants

▶ petroleum jelly

▶ measuring cup

▶ water

▶ water with nutrients

Draw Conclusions

Did your results support your hypothesis? Why or why not? Share your results with your classmates.

_____

_____

Open Inquiry

What other questions do you have about plants and their needs or structures? Talk with your classmates about questions you have. Choose one question to investigate. How might you answer this question? Make sure your experiment tests only one variable at a time.

My question is: _____

_____

How I can test it: _____

_____

My results are: _____

_____

_____

Name _____ Date _____

How do an animal's structures help it meet its needs?

Materials

- snail
- cotton swab
- water
- clear plastic container
- paper towel
- lettuce leaf

Purpose

Observe a snail to learn about its structures.

Procedure

1 **Observe** Look at the snail. What parts does it have? Do you see legs or eyes? Handle animals with care.

_____

_____

2 Draw the snail. Label all the parts you can.

3 **Predict** Which parts help the snail move? Which parts help it get food or stay safe?

_____

_____

_____

4 **Experiment** Gently touch the snail with a cotton swab. Observe the snail's actions for a few minutes. Record what you see.

Step **4**

_____

_____

5 **Experiment** Place a lettuce leaf in the container. Record the snail's actions.

_____

© Macmillan/McGraw-Hill

Explore

Draw Conclusions

6 **Communicate** On your drawing, circle the parts that the snail used to move and eat. Describe how it responded to its environment.

_____

_____

7 **Infer** Think about other animals you have seen, such as hamsters, birds, and fish. Do they have the same parts as the snail? Which parts do they use to meet their needs?

_____

_____

Explore More

Experiment Does the snail respond to light and dark? Make a plan and find out.

_____

_____

Open Inquiry

Design an activity to determine how a snail responds to another type of stimulus.

My question is: _____

_____

How I can test it: _____

_____

My results are: _____

Name _____ Date _____

How will a sow bug respond to its environment?

Make a Prediction

Look at a sow bug and predict how it will respond to a change in its environment.

_____

_____

Procedure

1 Observe Use the hand lens to look at the sow bug. What does it look like? Which parts does it have?

_____

_____

2 Predict What do you think will happen if you touch the sow bug with the tip of a cotton swab?

_____

_____

3 Experiment Gently touch the sow bug with the tip of a cotton swab.

4 Communicate Describe what happened to the sow bug when you touched it with the cotton swab. Was your prediction correct?

_____

_____

Observe Animal Structures

1 Look for photos of dolphins in magazines or on the Internet.

2 **Infer** Look at the dolphins' structures. How does a dolphin use its tail? How does it use its blowhole? What structures help it get food?

_____

_____

_____

3 **Communicate** Make a data table to show how each structure helps a dolphin meet its needs.

Body Part	Purpose
Sharp teeth	
Strong tail	
Fins	
Blowhole	

Classify

Earth is a big place. Millions of living things find homes in many different environments. With so many living things and so many environments, what can scientists do to understand life in our world? One thing they do is classify living things.

Learn It

When you classify, you put things into groups that are alike. Classifying is a useful tool for organizing and analyzing things. It is easier to study a few groups of things that are alike than millions of individual things.

© Macmillan/McGraw-Hill

Try It

Scientists classify plants. They classify animals, too.
Can you?

1 To start, observe the animals shown on page 51 of your student textbook. Look for things they have in common.

2 Then come up with a rule. What characteristic can you use to group the animals? Let's try wings. Which animals have wings? Which animals do not? Make a table to show your groups.

Wings	No Wings

Apply It

Classify these animals using your own rule.

How can you classify animals?

Purpose

Classify animals to form groups with similar characteristics.

Procedure

Step **1**

1 **Observe** Look at each animal. What structures does each animal have? Does each animal have legs? If so, how many? Does each animal have a distinct head and body?

_____

_____

_____

2 **Communicate** Make a chart like the one shown. Use words and pictures to describe characteristics of each animal.

3 **Classify** Put the animals into groups that are alike. Use the information in your chart to help you. Is there more than one way to group the animals?

_____

_____

Materials

- 4 plastic containers
- hand lens
- worm
- beetle
- snail
- ant

Step **2**

Animal Structure	beetle	snail	worm	ant
legs	6			
antennae	2			
head				
mouth				
eyes				
shell				

Name _____ Date _____

Draw Conclusions

4 **Interpret Data** Which two animals are most similar to
each other?

_____

_____

5 **Communicate** What rule did you use to classify the
animals? Why did you classify the animals the way
you did?

_____

_____

Explore More

Classify What other animals fit into your groups? Add
animals to each of your groups. Research any animals you
are not sure of.

_____

_____

Open Inquiry

Look at photos of several different animals. Can they be
classified?

My question is: _____

_____

How I can test it: _____

_____

My results are: _____

_____

How can you classify animals?

Material

• pictures of animals

Purpose

In this activity you will classify animals into groups with similar characteristics.

Procedure

1 **Observe** Look at all the pictures of the different animals. Arrange the animals into groups with the same characteristics.

2 **Communicate** What groups of animals do you have?

_____

_____

3 **Communicate** Fill in the chart below according to how you classified the animals.

Birds	Fish	Mammals

Draw Conclusions

4 **Communicate** Why did you classify the animals the way you did?

_____

_____

Name _____ Date _____

Model a Backbone

1 **Observe** Look at the photo of the raccoon on page 54 of your student textbook. What does its backbone look like?

_____

_____

_____

2 **Make a Model** Use clay and pipe cleaners to make a model of a backbone. Design your model so that it can bend from side to side and forward and backward.

3 **Experiment** How can your model move? Can you move one bone without moving all the others?

_____

_____

_____

4 **Infer** If your backbone were one solid bone, could it move as much?

_____

_____

_____

© Macmillan/McGraw-Hill

Name _____ Date _____

What does a seed need to grow?

Form a Hypothesis

Do seeds need water to grow? Form a hypothesis. Start with "If seeds do not get water, then . . ."

_____

_____

Test Your Hypothesis

1 **Observe** Look at the seeds with a hand lens. Draw what you see on a separate sheet of paper.

2 **Use Variables** Fold each paper towel into quarters. Then put two tablespoons of water onto one towel. Put the wet towel into a plastic bag. Label the bag *Water.* Put the dry towel into a bag. Label this bag *No Water.*

3 Place three seeds into each bag. Seal the bags and place them in a warm spot.

4 **Observe** Look at the seeds every day for a week. Record what you see with pictures and words. If the paper towel in the *Water* bag feels dry, add two tablespoons of water.

Materials

- 6 seeds

- 2 paper towels

- tablespoon

- hand lens

- water

- 2 plastic bags

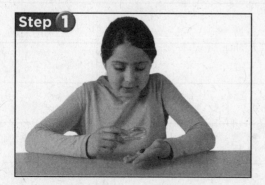

Step 1

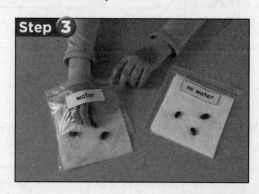

Step 3

© Macmillan/McGraw-Hill

Name _____ Date _____

Draw Conclusions

5 **Interpret Data** Which seeds changed? How did they change?

_____

_____

6 **Infer** Why do you think the seeds changed?

_____

Explore More

Experiment What would happen if you wet the paper towel with something other than water? Experiment to find out.

_____

_____

Open Inquiry

What other things do you think seeds need to sprout? Think of a question about what seeds need. Make a plan and carry out an experiment to answer your question.

My question is: _____

_____

How I can test it: _____

_____

_____

My results are: _____

_____

© Macmillan/McGraw-Hill

What else do seeds need in order to grow?

Materials

- encyclopedia
- computer

Make a Prediction

Besides water, predict what a seed will need in order to start growing.

_____

Test Your Prediction

1 **Research** Use your research materials to find instructions for growing three different types of plants. According to your research, what does a seed need in order to grow?

_____

Draw Conclusions

2 How good was your prediction?

_____

_____

3 **Think Critically** Which conditions were required for some seeds but not for others? Why do you think this is so?

_____

_____

_____

Name _____ Date _____

Fruits and Seeds

Materials

• fruit from 3 different plants

1 **Observe** Look at the fruits from three different plants. Compare their shapes and sizes.

2 Carefully cut open the fruits. How do their parts compare? Do they all have a peel or skin? Do they all have seeds?

_____

_____

_____

3 **Observe** Look at the seeds from each fruit. Compare the location of the seeds in each fruit.

4 **Infer** What do all fruits have in common? How might fruits help seeds survive and grow?

_____

_____

_____

_____

Form a Hypothesis

You just learned how seeds grow into plants. Can seeds grow in the cold? To answer questions like this, scientists start with what they know about plants. Then they use this information to turn their question into a testable statement. That is, they **form a hypothesis.**

Learn It

When you **form a hypothesis,** you make a statement that you can test by collecting data. Suppose you want to find out if plants need sunlight. Based on what you know, you could form a hypothesis like this: If plants do not get sunlight, then they will not grow.

A good hypothesis needs to be testable. You could test this statement by placing one plant in the dark and one in sunlight. Then you could observe and record what happens. A hypothesis also needs to identify the variables. In the example above, sunlight and plant growth are variables.

Name _____ Date _____

Try It

Form a hypothesis about what seeds need to grow. Then test that hypothesis with an experiment.

1 Think about what you know about seeds. Now form a hypothesis about this question. Will pea seeds germinate more quickly in a cold spot or in a warm spot? Begin with "If I plant a pea seed in the cold, then. . . ."

_____

_____

_____

2 Fold two wet paper towels in half, and place three seeds onto each. Place each paper towel into a plastic bag, and seal the bags.

3 Place one bag into a foam cup filled with ice. Place the other into an empty cup.

Step 3

© Macmillan/McGraw-Hill

4 Make a chart like the one below. Use it to record your observations. Do your results support your hypothesis?

_____

_____

Step 4	Cold	Warm
Day 1		
Day 2		
Day 3		
Day 4		

Apply It

Now that you have learned to think like a scientist, you can answer other questions. Do seeds germinate more quickly in the light or dark? **Form a hypothesis** about this question. Then plan an experiment to test your hypothesis.

_____

_____

_____

_____

_____

_____

Name _____ Date _____

How does a caterpillar grow and change?

Make a Prediction

How does a caterpillar change as it grows?
Write a prediction.

_____

_____

Test Your Prediction

1 **Observe** Look at the caterpillar. On a
separate piece of paper, draw a picture
of it and label all the parts you can see.
⚠ **Be Careful!** Handle animals with care.

2 **Measure** Find the length of your caterpillar. Record the
caterpillar's length on your drawing.

3 Put your caterpillar into the kit.

4 **Observe** Once a day, observe your caterpillar and draw
a picture of it. Label any changes you observe. If you
can measure the caterpillar's length without disturbing
it, record the length each day.

Materials

- caterpillar

- ruler

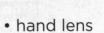

- hand lens

- caterpillar kit

Step **1**

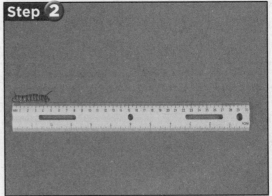

Step **2**

© Macmillan/McGraw-Hill

Draw Conclusions

5 **Interpret Data** What small changes did the caterpillar go through? What big changes did you observe?

_____

_____

6 **Infer** What are the stages in a butterfly's life cycle?

_____

Explore More

Experiment How do tadpoles change as they grow? Make a plan to test your ideas.

_____

_____

Open Inquiry

Explore another animal's growth. Think of a question about animal growth. Make a plan and carry out an experiment to answer the question.

My question is: _____

_____

How I can test it: _____

_____

My results are: _____

Name _____ Date _____

How do pets change as they grow?

Make a Prediction

Humans take care of many animals as pets. These animals live with us as they grow. How do the animals we choose as pets change as they grow older from birth to being adults?

_____

_____

_____

Draw Conclusions

1 Examine pictures of one or more animals that your classmates have as pets. Put the pictures of one animal in order, from youngest to oldest. Describe some of the changes you see.

_____

_____

_____

2 Was your prediction correct? What did you notice in the pictures that surprised you?

_____

_____

_____

A Bird's Life Cycle

1 **Observe** Look at these three photos. Put them in order to show the life cycle of a chicken.

_____

_____

2 **Communicate** Describe a chicken's life cycle. How does a chicken change as it grows?

_____

_____

_____

3 **Compare** How is the chicken's life cycle similar to the turtle's? How does it differ?

_____

_____

_____

Name _____ Date _____

Which traits are passed on from parents to their young?

Make a Prediction

Which of your traits are inherited, or passed on, from your parents? Is your hair color or hair length inherited? Write a prediction.

_____

_____

Test Your Prediction

1 **Communicate** Make a data table like the one shown. Use your table to describe your traits.

2 **Classify** Some traits have changed since you were little. Others have not changed. Circle the traits that have not changed.

Step 1	
Trait Name	**Trait Description**
Hair Color	
Hair Length	long/short (circle one)
Dimples	yes/no (circle one)
Ear Lobes	attached/unattached (circle one)
Favorite Food	

3 **Communicate** Compare tables with a classmate. Which of your classmate's traits have stayed the same over time?

_____

_____

Draw Conclusions

4 Which traits did most students classify as traits that stay the same?

_____

© Macmillan/McGraw-Hill

5 **Infer** How do you think you got the traits that do not change?

_____

6 **Infer** Some of your traits are inherited from your parents. Underline the traits that you think you inherited. Explain why you chose those traits.

_____

_____

Explore More

Make a trait table that has a column for each member of your family. Which traits do you share with your family members?

_____

_____

Open Inquiry

Think of a question about inherited and learned traits. Make a plan and carry out an experiment to answer your question.

My question is: _____

_____

How I can test it: _____

_____

My results are: _____

_____

What traits do people have?

Materials

• notebook

Make a Prediction

Take a moment to think about the characteristics, or traits, that make you similar to or different from your classmates. Then predict how your traits will differ from others' traits in your class. First, list some of your characteristics that you think will be different from your classmates.

_____

Test Your Prediction

❶ Take a few moments to look around and examine your classmates' traits. Which traits do you notice that you did not list above?

_____

_____

Draw Conclusions

❷ How are the traits that you noticed different among your classmates? How are their traits different from your traits? What traits do you think might be different that you cannot easily observe?

_____

_____

_____

© Macmillan/McGraw-Hill

Inherited Traits

1 **Observe** Look at the photo of the family on page 93 of your textbook. How are these people all alike? How do they differ?

_____

_____

2 **Communicate** Compare the children to their mother. Discuss which of their traits are similar to their mother's traits.

_____

_____

_____

3 **Communicate** Compare the children to their father. Discuss which of their traits are similar to their father's traits.

_____

_____

4 **Infer** Why do organisms look similar to, but not exactly like, their parents?

_____

_____

_____

Name _____ Date _____

What kind of food do owls need?

Materials

- plastic gloves
- paper plate
- owl pellet
- tweezers
- hand lens

Purpose

Find out what an owl eats by studying an owl pellet.

Procedure

1 Work with a partner. Put on plastic gloves. Place your owl pellet onto a paper plate.

2 **Predict** What do you expect to see inside the owl pellet? Write your prediction.

_____

_____

3 Using the tweezers, separate the objects in the owl pellet.

4 **Observe** What is in the owl pellet? Use the hand lens. Record your observations. ⚠ **Be Careful!** Wash your hands when you are done.

_____

_____

Step 3

© Macmillan/McGraw-Hill

Draw Conclusions

5 **Interpret Data** What do the materials inside the owl pellet tell you about what an owl eats?

_____

_____

6 **Infer** What organisms might an owl eat? What might those organisms eat?

_____

_____

_____

Explore More

Interpret Data Keep track of the things you eat in one day. Do most of your foods come from plants or animals?

_____

Open Inquiry

How do the diets of animals vary depending on the type of animal and where the animal lives? Think of your own question about the diet of animals. Make a plan and carry out an experiment to answer your question.

My question is: _____

_____

How I can test it: _____

_____

My results are: _____

_____

Where do the foods in your cafeteria come from?

Materials

• school cafeteria menu

Make a Prediction

Humans are omnivores, which means that they eat foods that come from both plants and animals. Predict where the foods you eat in the cafeteria come from.

_____

_____

Draw Conclusions

① **Research** Take a look at your school's cafeteria menu. List the foods. Next to each item, write where you think that food came from.

_____

_____

② **Draw Conclusions** Where did most of the cafeteria foods come from?

_____

_____

③ **Think Critically** Plants get their energy from the Sun. Where do the animals you eat get energy?

_____

_____

© Macmillan/McGraw-Hill

Observe Decomposers

1 Put some apple pieces into a plastic bag. Seal the bag.

⚠ **Be Careful!** Do not open the sealed bag.

2 **Observe** Leave the bag in a warm, dark place for a week. Observe the pieces every day. Record the changes you see.

3 **Communicate** What happened to the pieces of apple? How did they change over time?

_____

_____

_____

4 **Infer** What does this activity tell you about decomposers?

_____

_____

_____

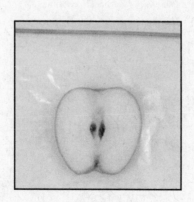

Name _____ Date _____

Communicate

You know that organisms get energy from food. Scientists study ecosystems to learn how different organisms get energy. Then they communicate, or share, their observations. Communicating helps people learn about the world.

Learn It

When you communicate, you share information with others. Some ways you share information in science are by talking, writing, drawing, or making graphs and charts.

Grassland Organisms	
Organism	**Where Organism Gets Energy**
grass	Sun
snake	field mouse
hawk	snake
field mouse	grass

© Macmillan/McGraw-Hill

Try It

In this activity you will organize and **communicate** data about a grassland ecosystem. Look at the data table on the previous page. It shows how some organisms in a grassland get energy. It also tells how the organisms interact. A table is one way to communicate data. You will try some other ways.

❶ One way you can communicate data is by making a food-chain diagram. The photographs on page 117 of your textbook show the start of a food-chain diagram. Part of the diagram is shown below. Complete it by adding the last three organisms in the correct order.

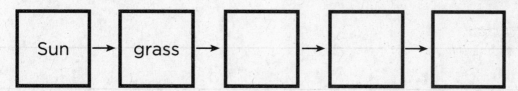

| Sun | → | grass | → | | → | | → | |

❷ Next, communicate by making a food pyramid. Fill in the blank spaces on the pyramid above.

❸ Now, communicate by writing a paragraph. In your paragraph, classify each organism as a producer or consumer. Tell where each grassland organism gets its energy.

❹ Did all three ways of communicating help you understand the data? Which way did you think worked best? Why?

_____

_____

Apply It

Think of a food chain from another ecosystem.
Communicate information about this food chain to a
partner. Draw a food-chain diagram to show where
organisms in the ecosystem get energy. Now describe
the food chain in words. Discuss what you learned.

_____

_____

_____

_____

© Macmillan/McGraw-Hill

Can ocean animals live and grow in fresh water?

Make a Prediction

Can brine shrimp grow in both fresh water and salt water? Write a prediction.

_____

_____

Test Your Prediction

1 Fill each jar with 480 mL of water. Put two tablespoons of sea salt in one jar. Label it *Salt Water*. Label the other jar *Fresh Water*.

2 Add one teaspoon of brine shrimp eggs to each jar.

3 **Observe** Watch what develops in each jar over the next few days. Use a hand lens.

_____

_____

_____

Materials

• 2 jars

• measuring cup and water

• sea salt

• measuring spoon

• brine shrimp eggs

• hand lens

Step 1

Salt Water Fresh Water

Draw Conclusions

4 **Interpret Data** In which jar did the brine shrimp eggs hatch? How could you tell?

_____

5 **Infer** Can all ocean animals live and grow in fresh water? How do you know?

_____

_____

_____

Explore More

Experiment Does temperature affect the hatching of brine shrimp eggs? Design an experiment to find out.

_____

_____

Open Inquiry

How would the type of water used for watering a plant affect the plant's growth? Think of your own question about plants and how they grow. Make a plan and carry out an experiment to answer your question.

My question is: _____

_____

How I can test it: _____

_____

My results are: _____

© Macmillan/McGraw-Hill

How are desert and forest plants different?

Materials
• cactus plant
• fern plant

Make a Prediction

Plants and animals have different characteristics
that allow them to live in different types of environments.
Predict which characteristics allow cactus plants and fern
plants to live in different ecosystems.

_____

_____

Draw Conclusions

1 **Observe** Look at the cactus plant and the fern plant
side by side. How are their stems and leaves different?

_____

_____

_____

2 How do you think these features are helpful in a desert
or forest environment?

_____

_____

_____

_____

_____

Name _____ Date _____

Water Temperatures

1 Fill two jars each with 200 mL of salt water. Label one jar *Sunlight* and put it in a sunny place. Label the other jar *No Sunlight* and put it in a very dark place.

2 **Observe** Measure the water temperature in each jar with a thermometer later in the day. Which jar is warmer?

_____

3 **Infer** The two jars model two parts of the ocean. What are those parts? How are they different?

_____

_____

_____

_____

© Macmillan/McGraw-Hill

Does fat help animals survive in cold environments?

Materials

- vegetable fat
- paper towel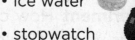
- ice water
- stopwatch

Form a Hypothesis

Can fat help keep your finger warm in cold water? Write a hypothesis. Write your answer in the form "If my finger has a layer of fat, then . . ."

_____

_____

Test Your Hypothesis

1. Use a paper towel to spread vegetable fat over one index finger. Try to coat it completely. Leave your other index finger uncovered.

2. **Predict** What will happen when you put both index fingers in a bowl of ice water?

_____

3. **Experiment** Put one index finger into the ice water. Ask a partner to time how long you can keep your finger in the water. Repeat with your other index finger. Record the data in a chart on a separate piece of paper.

4. Trade roles with your partner and repeat steps 1 through 3.

Draw Conclusions

5. **Interpret Data** Which finger could you keep in the ice water longer? Why? Did your results support your hypothesis?

_____

6 **Infer** Walruses have a layer of fat under their skin. How does this help them survive?

_____

Explore More

Experiment How could you measure how well fat keeps things warm? Could you use thermometers? Make a plan and test it.

_____

_____

Open Inquiry

What adaptations do animals have that allow them to survive comfortably in hot weather? Think of your own question about animal adaptations. Make a plan and carry out an experiment to answer your question.

My question is: _____

_____

How I can test it: _____

_____

My results are: _____

_____

What adaptations can you observe?

Make a Prediction

All animals have adaptations that help them
live in certain environments. Predict one feature of an
animal that helps it adapt to its environment. How does the
adaptation help the animal live in a particular environment?

_____

_____

Materials
• photos of various animals

Draw Conclusions

1 **Observe** Look at the pictures of animals that your
teacher has provided. List some of the adaptations you
see in the photographs.

_____

_____

_____

2 **Interpret Data** Describe how each adaptation helps an
animal live in a certain environment.

_____

_____

_____

_____

Name _____ Date _____

Storing Water

1 **Make a Model** Wet two paper towels. Then wrap one in wax paper. This models a plant that has waxy skin. Use the uncovered towel to model a plant that does not have waxy skin.

2 Place your models in a sunny window.

3 **Observe** How do the paper towels feel later in the day?

_____

_____

_____

4 **Infer** How does waxy skin help desert plants survive?

_____

_____

_____

Structured Inquiry

How does camouflage help some animals stay safe?

Form a Hypothesis

Which is easier to find, an animal that blends into its environment or an animal that does not blend in? Form a hypothesis. Start with "If an animal blends into its environment, then . . ."

_____

_____

Test Your Hypothesis

1 Cut out 20 yellow circles and 20 brown circles.

2 **Experiment** Spread out the circles on yellow paper to model animals with and without camouflage. Then ask a classmate to pick up as many circles as he or she can in 10 seconds.

Materials

- yellow paper

- brown paper

- scissors

- stopwatch

Step **2**

© Macmillan/McGraw-Hill

3 **Communicate** How many of each color circle did your classmate pick up? In the space below, create a chart like the one shown in order to record the results.

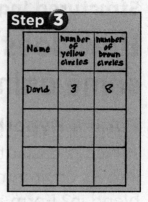

Step **3**

Name	number of yellow circles	number of brown circles
David	3	8

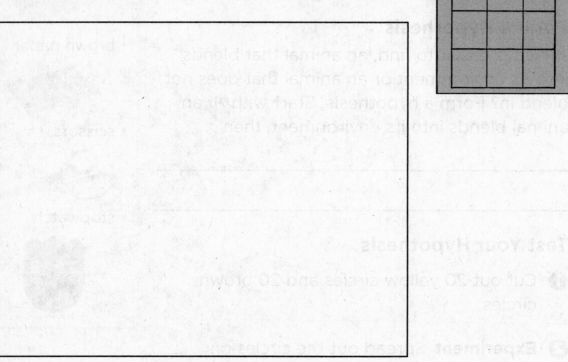

4 Repeat steps 1 and 2 with two other classmates.

Draw Conclusions

5 **Interpret Data** Did your classmates pick up more yellow or brown circles? Which circles were harder to find?

_____

_____

6 **Infer** How does camouflage help animals stay safe?

_____

_____

© Macmillan/McGraw-Hill

Name _____ Date _____

Guided Inquiry

How do pale colors help some animals survive?

Form a Hypothesis

How do pale body coverings affect a desert animal's temperature? Write a hypothesis.

_____

_____

_____

Materials

- black beans
- white beans
- 2 thermometers

Test Your Hypothesis

Design a plan to test your hypothesis. Use the materials shown. Write the steps you plan to follow.

_____

_____

_____

_____

_____

Draw Conclusions

Did your results support your hypothesis? Why or why not? Share your results with your classmates.

_____

_____

_____

Open Inquiry

What other questions do you have about plant and animal adaptations? Discuss with classmates the questions you have. How might you find the answers to your questions?

My question is: _____

_____

How I can test it: _____

_____

My results are: _____

_____

© Macmillan/McGraw-Hill

How can worms change their environment?

Purpose

All living things change their environments as they get food, water, shelter, and other needs. In this activity, find out how worms change their environment.

Procedure

1. **Make a Model** Put some soil in a plastic container. Then put small stones and leaves on top of the soil. This models a forest floor.

Step 1

2. Place live worms on the "forest floor."

3. **Predict** What will the worms do? Make a short list of the things you might see the worms do.

Step 2

_____

_____

_____

4. **Observe** Check the worms, soil, leaves, and stones every three to four days. Keep the soil moist. Record your observations.

Draw Conclusions

5 **Infer** What happened to the leaves over time?

_____

6 **Communicate** How do worms change the environment in which they live?

_____

Explore More

Experiment How do other living things change their environments? Make a plan to test your ideas. Then try your plan.

_____

_____

Open Inquiry

How can you measure the changes that an organism makes to an environment? Think of your own question about how changes to the environment can be measured. Make a plan and carry out an experiment to answer your question.

My question is: _____

_____

How I can test it: _____

_____

My results are: _____

_____

How does a forest form?

Purpose

The forests you hike through took several hundreds of years to form. The land requires many years of growth and change before it becomes a forest. At each stage, the land looks a little different. This is because the plants that grow change the land in stages and remake the environment for other plants to grow in the next stage.

Materials

• books that show the different phases of forest formation

Procedure

❶ Study the illustrations of a forest-forming sequence. What types of plants live on the land in the beginning?

_____

❷ How do the types of plants living on the land change through the years?

_____

_____

Draw Conclusions

❸ What do you think happens to young plants when they die? How might this change the soil in a way that helps older plants grow?

_____

_____

_____

Name _____ Date _____

Model Pollution

1 **Observe** Look at the shell of a hard-boiled egg. Is it hard or soft? Why do you think the egg has this type of shell?

_____

_____

2 **Make a Model** Fill a large cup with vinegar. This models polluted land or water. Place your egg inside the cup.

3 **Observe** Look at the egg throughout the day. Study the shell of the egg. Do you notice any differences in the egg or its shell?

_____

_____

_____

_____

4 **Infer** After being placed in vinegar, can the shell still protect the egg?

_____

_____

5 **Predict** What may happen to eggs near polluted land or water?

_____

_____

© Macmillan/McGraw-Hill

Use Numbers

The average American changes his or her environment by producing about 2 kg (4 lbs) of trash every day. We can never get rid of trash completely. However, we can cut down the amount we create by practicing the 3 Rs. Do students in your school practice the 3 Rs? Find out the same way scientists do: **use numbers** to record data.

Learn It

When you **use numbers,** you present data in a way that people can clearly understand. Basic math skills, such as counting and ordering numbers, help to collect and organize information. Often, scientists gather and record data by asking questions or by having people fill out surveys. Then they use numbers to put the data into a chart or graph. You can do it, too.

Try It

In this activity, you will gather data and **use numbers** to find out how much trash is thrown out by students in your school. You cannot survey the whole school, but you can do a mini-survey.

1 Choose five students to survey in the lunchroom.

2 Ask each student questions about how many pieces of trash from lunch he or she threw away today. Ask about the containers used. Will anything be reused?

3 Use a table like the one shown on the next page to organize your data.

Now use numbers to answer the following questions:

▶ Did every student throw out some trash or packaging material?

_____

_____

▶ How many pieces of trash did the students recycle? How many pieces did they reuse?

_____

_____

▶ How many pieces of trash did these five people create altogether?

_____

_____

Student's Name	Pieces Reused	Pieces Recycled	Pieces Thrown Away	Total Pieces of Trash
Total				

Apply It

Use numbers to combine your data with those of your classmates. Add to find the total for each column. Then make a bar graph to show the results.

Do you predict these same students will throw out more or less trash tomorrow? Plan another survey. Then use numbers to compare the new results to your first results just as scientists do!

_____

_____

Name _____ Date _____

How can a flood affect plants?

Form a Hypothesis

What happens to plants when they get too much water? Write a hypothesis.

_____

_____

Test Your Hypothesis

1 Label three plants *A*, *B*, and *C*. Water plant *A* once a week with 60 mL of water. Water plant *B* every day with 60 mL of water. Water plant *C* every day with 120 mL of water.

Step 1

2 **Predict** Which plant will grow to be the tallest? Record your prediction.

_____

_____

3 **Observe** Monitor your plants every few days. Measure how tall they grow. Record how they look with words and pictures on a separate piece of paper.

Step 3

© Macmillan/McGraw-Hill

Draw Conclusions

④ Interpret Data How did the plants change over time? Which plant grew the tallest? Which do you think is the healthiest?

_____

⑤ Infer What happens to some plants when there is a flood?

_____

_____

Explore More

Experiment Could your plant recover from a flood? Stop watering plant *C* for a week. How does the plant change?

_____

_____

Open Inquiry

In what other ways does a plant's environment change? Think of your own question about changes to a plant's environment. Make a plan and carry out an experiment to answer your question.

My question is: _____

_____

How I can test it: _____

_____

My results are: _____

_____

Name _____ Date _____

How do plants respond to temperature changes?

Materials
- two small fern plants
- thermometer
- water

Form a Hypothesis

Different plants grow with different amounts of water available and different temperatures. Fern plants typically grow in warm, humid environments. How do you think a fern plant would respond if the environment suddenly became very cold?

_____

_____

Test Your Hypothesis

1. Place one fern plant in a shaded spot inside the classroom. Place another fern plant outside where temperatures are at least 5–10 degrees Celsius lower than the classroom temperature. Water each plant with the same amount of water.

2. Observe the plants every day for the next few days. What do you notice?

_____

_____

Draw Conclusions

3. How did the temperature change affect the plant that was placed outside? Would a fern plant survive if the climate suddenly turned very cold?

_____

_____

A Changing Ecosystem

1 Make five character cards. Label the cards: prairie dog, snake, burrowing owl, eagle, and coyote.

2 Paste the cards on a large sheet of paper.

3 Draw an arrow from each animal to the organisms it depends upon for food or shelter.

4 **Infer** What would happen if prairie dogs disappeared?

_____

_____

_____

5 **Infer** What would happen if eagles disappeared?

_____

_____

_____

snake

Name _____ Date _____

How do fossils tell us about the past?

Purpose

Find out how fossils can teach about the past.

Procedure

Materials

- measuring cup and water
- glue
- colored sand
- paper cup
- "fossil" objects
- brush

① Mix a little glue and water in a measuring cup.

② **Make a Model** Pour a thin layer of colored sand into a paper cup. Add a "fossil" object. Cover the object with sand of the same color. Add a little water and glue to "set" this layer. This models a fossil in rock.

③ Repeat step 2 with different objects and different colors of sand. Make three layers in all. Allow the layers to dry.

④ **Observe** Trade cups with another group. Carefully peel the paper cup away. Use the brush to find the fossils. Start at the top layer. Work your way down.

Step 4

⑤ **Communicate** Record in a table the order in which each fossil object was found.

Step 5 Layer	Fossil
top	
middle	
bottom	

Draw Conclusions

6 **Interpret Data** Which fossil was buried first? Last? Which fossil is oldest?

_____

7 **Infer** What can layers of rock tell us about Earth's past?

_____

_____

Explore More

How else could you model a fossil? Make a plan and try it.

_____

Open Inquiry

How does the material from which fossils are formed affect the condition they are found in? Think of your own question about fossil formation. Make a plan and carry out an experiment to answer your question.

My question is: _____

_____

How I can test it: _____

_____

_____

My results are: _____

_____

What can you learn from a fossil?

Materials
• trilobite fossil

Purpose

Fossils are the preserved remains of organisms that lived long ago. What types of things might you be able to learn from looking at a fossil?

_____

_____

Procedure

1 Observe the fossil sample supplied by your teacher. Describe the fossil.

_____

_____

2 Do you think that this fossil comes from a plant or an animal?

_____

_____

3 Does this fossil remind you of any modern-day organisms?

_____

_____

Fossil Mystery

1 **Make a Model** Choose your favorite animal. Then use the key below to make fossil marks for your animal on some modeling clay.

If your animal is a . . .	then shape the clay into a . . .
mammal	circle
bird	square
amphibian	rectangle
reptile	triangle
fish	ball

2 Use the key below to make more fossil marks.

If your animal . . .	then mark your clay with . . .
lives in water	fins
lives on land	feet
lives both in water and on land	fins and feet
is a carnivore	pointed teeth
is an herbivore	flat teeth
is an omnivore	pointed and flat teeth

3 Trade your model fossil with the person sitting to your right.

4 **Infer** What can you learn about the animal that your classmate chose? How do scientists use fossils to learn about extinct animals?

_____

_____

_____

_____

Name _____ Date _____

Does land or water cover more of Earth's surface?

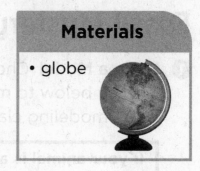

Make a Prediction

Do you think that there is more land or more water on Earth's surface? Write your prediction.

_____

_____

Test your Prediction

① Make a table like the one shown for 10 spins.

② **Experiment** Slowly spin a globe. Do not look at it. Touch your finger to the globe to stop it.

Step **1**		
Spin	**Land**	**Water**
1		
2		
3		
4		

③ **Observe** Did your finger stop on land or water? Record the information on the chart.

_____

_____

Step **2**

④ Repeat steps 2 and 3 nine more times.

⑤ **Use Numbers** How many times did you touch water? How many times did you touch land?

_____

© Macmillan/McGraw-Hill

Draw Conclusions

6 **Infer** Is there more land or more water on Earth? How do your results compare with the results of others?

_____

_____

Explore More

Experiment Which covers more of Earth, rivers or oceans? Make a plan to find out.

_____

_____

Open Inquiry

Determine which ocean covers the largest area on Earth's surface.

My question is: _____

_____

How I can test it: _____

_____

My results are: _____

_____

Name _____ Date _____

What are some land features?

Purpose

Investigate different types of features on land.

Procedure

1 **Observe** Use the globe, atlases, or maps to identify as many different landforms as you can. List the different landforms you found.

_____

_____

_____

2 Describe the different landforms.

_____

_____

_____

_____

_____

_____

Materials

- globe
- atlases
- physiographic maps

Your State's Features

1 **Make a Model** Draw a map of your state. Decide how to show your state's land and water features. Then make a key and complete the map.

2 **Observe** Where is your town or city located? Draw a large dot there. Which landforms and water features are found in your town or city? How do these features compare with those found in other parts of your state?

_____

_____

_____

_____

_____

_____

Make a Model

You just learned about many landforms. Some of them are found on land. Some lie under the ocean. In some places a limestone cave forms below the ground. It forms when water seeps into the ground and changes rock. This can take millions of years. You can **make a model** to show a cave.

Learn It

When you **make a model**, you build something to represent, or stand for, a real object or event. A model may be bigger or smaller than the real thing. Models help you learn about objects or events that are hard to observe directly. Maps and globes are two examples of models.

Try It

In this activity, you will **make a model** of a cave.

Materials

- ruler
- scissors
- tan or white construction paper
- crayon
- shoe box or other small box
- clear tape

1 Cut a piece of construction paper so that it is a little smaller than the size of the back wall of the box.

2 On the paper draw limestone rocks like the ones shown on page 201 of your student textbook. Tape the paper to the box's back wall.

3 Draw more limestone rocks on another piece of construction paper. Draw a flap for each rock.

4 Cut out each rock and its flap. Bend the flap for each rock. Tape each rock inside the box. Use the photo of the model to help you.

© Macmillan/McGraw-Hill

Now use your model to answer these questions:

▶ How would you describe the shapes of rocks in a limestone cave?

_____

▶ Where do the rocks form?

_____

Apply It

Make a model of a landform that you learned about in this lesson. It may be a landform on the ocean floor or one above the ocean. What details do you want to show? Which materials will you use to help you model these details?

_____

_____

_____

_____

_____

_____

How does sudden movement change the land?

Purpose

Model what happens when the land suddenly moves.

Procedure

① **Make a Model** Fill a pan halfway with sand. Form a mountain in the sand.

② Place blocks in the sand to model buildings. Add twigs to model trees.

③ **Communicate** Draw your land surface.

④ **Experiment** What will happen if you tap the pan gently? Try it.

_____

_____

⑤ **Experiment** What will happen if you tap the pan harder? Try it.

_____

_____

Materials

- aluminum pan

- sand

- assorted blocks

- twigs

Step 2

Draw Conclusions

6 **Infer** How can the sudden movement of land change the land?

_____

_____

Explore More

Experiment Different rocks and soils make up land. Does sudden movement change all land the same way? Make a plan to find out. Then try it.

_____

_____

Open Inquiry

Think about whether liquid material will be affected differently by sudden movements of the land. Form a hypothesis and design an experiment to test it.

My question is: _____

_____

How I can test it: _____

_____

My results are: _____

_____

_____

© Macmillan/McGraw-Hill

How are buildings affected by movement?

Materials
- small blocks
- glue
- toothpicks
- rubber bands

Purpose

To model what happens to buildings during an earthquake.

Procedure

1 **Make a Model** Build three model buildings; one using just blocks, the second using blocks held together by the rubber bands, and the third by gluing toothpicks together to form a building.

2 **Measure** Place each model building on a desk and gently hit the edge of the desk to simulate an earthquake. Observe how each building is affected by the movement.

_____

_____

_____

Draw Conclusions

3 How is the way a building is made related to the amount of damage it might have from an earthquake?

_____

_____

Name _____ Date _____

A Model Volcano

1 **Make a Model** Cover a desk with newspaper. Place a small tube of toothpaste on the desk to model Earth's surface.

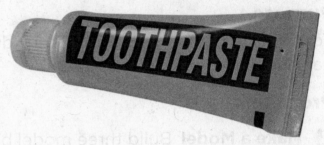

2 Carefully make a hole in the tube on the side opposite the cap. This represents an opening in Earth's surface.

3 **Observe** Press on the tube near the cap. What happens by the hole? What do you think the toothpaste is a model of?

_____

_____

_____

4 **Communicate** Did the same thing happen to everyone's tube? What was different? Why were there differences?

_____

_____

_____

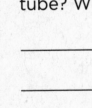

How can rocks change in moving water?

Form a Hypothesis

What happens to rocks when they move around in water? Write a hypothesis in the form, "If I shake rocks in water, then . . ."

_____

_____

Materials

- sandstone rocks

- measuring cup

- 3 plastic jars with lids

- stopwatch

- hand lens

Test Your Hypothesis

Step 1

❶ **Measure** Label the jars *A*, *B*, and *C*. Put the same number of similar-size rocks in each jar. Using the measuring cup, fill each jar with the same amount of water. Put a lid on each jar.

❷ Let jar *A* sit. Do not shake it.

❸ **Use Variables** Shake jar *B* hard for 2 minutes. Then let the jar sit.

Step 3

❹ **Use Variables** Shake jar *C* hard for 5 minutes. Then let the jar sit.

❺ **Observe** Use a hand lens to observe the rocks in each jar. What happened? Did the results support your hypothesis?

_____

_____

_____

Name _____ Date _____

Draw Conclusions

6 **Infer** How can rocks change in moving water?

_____

_____

Explore More

Experiment Would the results be the same if different rocks were used? Make a plan and try it.

_____

_____

_____

Open Inquiry

Think about whether rocks would break more easily without water in the jars. Formulate a question on this topic and then design and carry out an experiment to answer it.

My question is: _____

_____

How I can test it: _____

_____

My results are: _____

_____

_____

© Macmillan/McGraw-Hill

Do all rocks weather at the same rate?

Materials

- plastic jar with lid
- small sandstone or shale pieces
- granite pieces or marbles
- water

Form a Hypothesis

Form a hypothesis about whether all rocks weather at the same rate.

_____

_____

_____

Test Your Hypothesis

1. **Investigate** Place some softer rock pieces, like sandstone or shale, in the jar with some harder rock pieces, like granite or marbles. Add some water and tightly close the jar. Shake the jar quickly for five minutes.

2. **Predict** Which rocks do you think will be weathered more?

_____

_____

3. **Observe** Which type of rock was weathered the most?

_____

Draw Conclusions

4. Was your hypothesis supported by the results?

_____

_____

© Macmillan/McGraw-Hill

Name _____ Date _____

Materials Settle

1 **Make a Model** Pour one cup each of sand, soil, and pebbles into a jar. Fill the jar almost to the top with water. Seal the jar tightly.

2 Shake the jar 10 times. Then let it sit. Draw what you see on a separate sheet of paper.

3 **Interpret Data** In which order do the materials settle?

_____

_____

_____

4 **Infer** What happens to eroded materials in a river as the river gradually slows down?

_____

_____

_____

_____

How do a mineral's color and mark compare?

Materials

• minerals

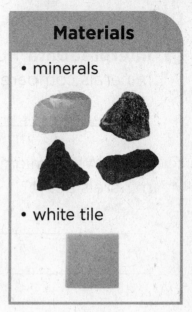

• white tile

Make a Prediction

Some minerals leave a mark behind when you rub them on a white tile. Is the mark left behind always the same color as the mineral?

_____

_____

_____

Test Your Prediction

1 Make a table like the one shown.

Mineral Color	Color Left Behind

2 **Observe** Look at one mineral. Record its color in the table.

3 **Experiment** Rub the mineral across the tile. What color is left behind? Record the color in the table.

4 Repeat steps 2 and 3 for each mineral.

Step 3

Name _____ Date _____

Draw Conclusions

5 **Interpret Data** How did the colors and marks of the minerals compare?

_____

6 **Infer** When might you use mineral marks to help tell minerals apart?

_____

_____

Explore More
Experiment Are some minerals harder than others? Make a plan to find out. Then try it.

_____

_____

Open Inquiry
Most rocks are made up of two or more minerals. How would you identify all the minerals in such a rock? Think of your own question about identifying minerals. Make a plan and carry out an experiment to answer your question.

My question is: _____

_____

How I can test it: _____

_____

My results are: _____

_____

How does a mineral break?

Purpose

Observe how to identify a mineral by how it breaks.

Materials

- table salt
- mica
- hand lens

Procedure

1 **Observe** Place a few grains of salt on a dark surface. Use a hand lens to look at the grains. What is the shape of the grains?

_____

2 **Predict** Do you think you could break a grain of salt into smaller pieces? What shape do you think the smaller pieces would have?

_____

3 **Observe** Use the hand lens to look at the mica. How does the mica break?

_____

4 **Infer** Do you think the way a mineral breaks can be used to identify it? Explain your answer.

_____

_____

_____

_____

Name _____ Date _____

Classify Rocks

1 **Observe** Use a hand lens to observe a few igneous rocks. What color are they? Are their grains large or small? Do they have a coarse texture or a fine texture?

_____

_____

_____

_____

2 **Classify** Put the rocks into groups that are alike.

3 **Infer** Which of the rocks do you think formed below Earth's surface? Which formed above Earth's surface? Explain why.

_____

_____

_____

_____

_____

© Macmillan/McGraw-Hill

Name _____ Date _____

What makes up soil?

Purpose
Find out what soil is made of.

Procedure

1 Use a spoon to spread out the soil on the plate.

2 **Observe** Use the hand lens to observe the soil. Is soil made of small bits of stuff? What is the shape and color of these small particles? Wash your hands. Record what you see.

_____

_____

_____

_____

3 **Communicate** Talk with others about what the tiny bits in soil may be.

Draw Conclusions

4 **Infer** What kinds of things make up soil?

_____

_____

_____

Materials

- plastic spoon

- soil

- paper plate

- hand lens

Step **1**

Step **2**

Name _____ Date _____

Explore More

Experiment Is all soil the same? Make a plan to find out.
Then try out your plan.

_____

_____

_____

Open Inquiry

What part of soil helps plants grow? Think of your own
question about what plants need from soil. Make a plan and
carry out an experiment to answer your question.

My question is: _____

_____

How I can test it: _____

_____

_____

My results are: _____

_____

© Macmillan/McGraw-Hill

Can water separate soil parts?

Materials
- soil
- jar
- hand lens

Purpose
Describe how water can be used to separate a mixture.

Procedure

1 Put soil into the jar until it is about 1 inch deep. Add water to the jar until it is about half full.

2 **Observe** Carefully swirl the contents of the jar for 1 minute. If the jar has a lid, you can carefully shake the jar and its contents. What do you observe in the jar?

_____

_____

3 **Observe** Let the jar sit still until the water looks almost clear. What happened to the soil?

_____

_____

4 **Communicate** Draw what you see in the jar. Label what you see in each layer. Use a hand lens to look at the soil if you need to.

5 **Conclude** Could water be used to separate the parts of some mixtures? Explain your answer.

_____

_____

Name _____ Date _____

Classify Soils

1 **Observe** Look at the two soils in plastic bags. How are they alike? How are they different?

_____

_____

_____

2 **Observe** Use a hand lens to look closely at each soil. Which soil has larger grains?

_____

_____

3 **Classify** Which soil is sandy soil? Which is clay soil? How do you know?

_____

_____

© Macmillan/McGraw-Hill

Name _____ Date _____

Use Variables

Soils differ from place to place. They contain different amounts of humus and are made up of different kinds of rocks. Do all soils hold the same amount of water? To answer this question, you can **use variables** to test how water moves through different soils.

Learn It

When you **use variables,** you identify things in an experiment that can be changed. Soil type is a variable, for example. The amount of soil you use in an experiment is also a variable. It is important that you change only one variable at a time when you experiment. You should keep all other variables the same. That way you can tell what caused the results.

© Macmillan/McGraw-Hill

Name _____ Date _____

Try It

You will **use variables** to answer this question: Does sandy soil or potting soil hold more water?

Materials

- pencil
- 4 disposable cups
- 250 mL of potting soil
- measuring cup
- water
- 250 mL of sandy soil
- watch or clock

1 Use a pencil point to poke three tiny holes in the bottom of a cup.

2 Put 250 mL of potting soil into the cup. Pack the soil firmly.

3 Fill a measuring cup with 100 mL of water.

4 Hold the cup of potting soil over an empty cup without holes. Slowly pour the water over the soil. Wait for two minutes. Write your observations in a table like the one shown on the next page.

5 Pour the water that drained out into the measuring cup. Record the volume in your table.

6 Repeat steps 1–5 using sandy soil in place of potting soil. Record the results.

7 Which soil held more water? How did changing the variable change the results?

_____

_____

_____

Variable	My Observations	Volume that Drained

Apply It

Now **use variables** to experiment more. Choose one of the following variables to test. List the variable in a table and record the results of your experiment. Did changing the variable change the results? If so, how?

▶ Do not pack the potting soil firmly.

▶ Mix clay into the sandy soil.

▶ Mix larger rocks into the potting soil.

▶ Poke larger holes in the cups.

_____

_____

_____

_____

How do some fossils form?

Purpose

Find out how some living things of the past become fossils.

Procedure

1. **Make a Model** Hold a spoon over a paper towel. Squeeze a small amount of glue onto the spoon. Let the glue set for 10 minutes. This models sticky tree resin.

2. **Make a Model** Place a thin apple slice on top of the glue. This models an organism trapped in tree resin. Slowly add more glue until the apple slice is completely covered.

3. **Use Variables** Put the spoon on a paper towel. Place the other apple slice next to the spoon.

4. **Observe** Look at the apple slices throughout the day. Record any changes you observe.

Draw Conclusions

5. **Interpret Data** Compare the two apple slices. What differences do you notice?

_____

_____

Materials

- plastic spoon
- paper towel
- glue
- 2 apple slices

Step 2

© Macmillan/McGraw-Hill

6 **Infer** What caused any differences you observed?

_____

7 **Infer** How do some fossils form?

_____

Explore More

Experiment Could an organism become a fossil in ice?
Make a plan to find out.

_____

_____

_____

Open Inquiry

Would a jellyfish make a good fossil? How about a fish?
Think of your own question about what kinds of things
make good fossils. Make a plan and carry out an experiment
to answer your question.

My question is: _____

_____

How I can test it: _____

_____

_____

My results are: _____

_____

How can you model forming a "fossil"?

Materials

• plaster of paris

• marble

• small paper cup

• petroleum jelly

Purpose

Model how one type of fossil is formed.

Procedure

1. Completely coat a marble with a little bit of petroleum jelly. You will need the marble in step 3.

2. Fill a paper cup half way with plaster of paris. You must use the plaster quickly before it hardens.

3. Press the marble into the plaster so that the marble is half in the plaster. Let the cup sit until the plaster hardens.

4. After the plaster hardens, coat the top of it with a thin coat of petroleum jelly. Then add about an inch more plaster to the cup. Let this plaster harden also.

5. **Observe** Carefully tear away the paper cup. Pull the top piece of plaster away from the bottom piece. Remove the marble from the plaster. What do you observe?

_____

6. **Compare** How did your experiment model how a fossil forms?

_____

_____

_____

© Macmillan/McGraw-Hill

Model Imprints

1 Break a small chunk of clay into two pieces. Roll each piece to form a ball.

2 **Make a Model** Take one clay ball. Press the front of your thumb into it. Take the other clay ball. Press the back of your thumb into it.

3 **Communicate** Switch clay balls with someone. How are the imprints like yours? How are they different?

_____

_____

_____

4 **Infer** What can you learn by comparing fossil imprints?

_____

_____

_____

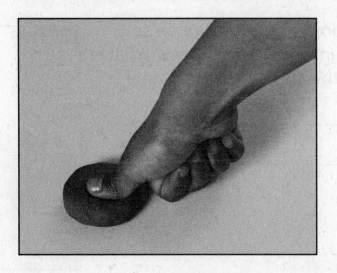

© Macmillan/McGraw-Hill

Name _____ Date _____

How is Earth's water made clean?

Purpose
Find out how Earth's water can be made clean.

Procedure

1 **Make a Model** Place a funnel inside a large, clear plastic cup. Use a spoon to fill the funnel with gravel half way. Then fill the rest of the funnel with coarse sand. What do you think these layers model?

Step 1

_____

_____

2 **Observe** Mix a little soil and some crushed leaves into a small, clear cup of water. On a separate piece of paper, draw what you see.

3 **Observe** Slowly pour the water, soil, and leaves into the funnel. Draw what you see in the cup under the funnel.

Step 3

_____

_____

Materials

• funnel

• spoon and gravel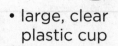

• clear cup of water

• large, clear plastic cup

• coarse sand

• soil and crushed leaves

Draw Conclusions

4 How did the water in step 3 compare with the water in step 2?

_____

_____

5 How can Earth's water be made clean?

_____

_____

Explore More

Experiment Can food coloring be removed from water?
Repeat the activity to find out.

_____

_____

Open Inquiry

What is the best way to get water clean? Think of your own
questions about what could be used to get water cleaner. Make
a plan and carry out an experiment to answer your question.

My question is: _____

_____

How I can test it: _____

_____

_____

My results are: _____

_____

How does weather clean water?

Purpose

Show how Earth's water cycle cleans dirty water.

Procedure

1 Pour very warm water into the cup until it is about half full.

2 Add a spoonful of soil to the water and stir.

3 Place the saucer over the top of the cup. Put an ice cube on the saucer.

4 **Observe** Let the cup and saucer sit for 5 minutes. Then look at the bottom of the saucer. Draw and label what you see.

5 **Communicate** What did you see on the bottom of the saucer?

_____

_____

6 **Compare** How was the water on the saucer and the water in the bottom of the cup different?

_____

_____

_____

Materials
• ice
• saucer
• warm water
• clear cup
• soil
• spoon

© Macmillan/McGraw-Hill

Record Water Use

1 Communicate Make a table like the one shown. Record each activity in which you use water.

Activities	Water Use

2 Use Numbers Use the table below to see how much water each activity uses. Record the amounts in your table.

Ways People Use Water	
Activity	**Normal Daily Use**
showering	
bathing	
brushing teeth	
washing hands	
running a dishwasher	
washing clothes	

3 Use Numbers How much water do you use each day? Do you use a lot of water? How could you use less water?

_____

_____

_____

Structured Inquiry

What things pollute the air?

Form a Hypothesis

What things are carried in air? Are the same things found in air everywhere? Record your hypothesis. Start with "If I hang white papers covered with petroleum jelly in several locations, then . . ."

_____

_____

_____

Materials

- 3 sheets of white paper

- craft stick

- petroleum jelly

- 3 pieces of string

- hand lens

Test Your Hypothesis

1 **Use Variables** Choose different locations to hang three sheets of white paper. Some examples are near a heating or cooling vent, near grass or trees, and near a sidewalk. Write your name and the location on the bottom of each paper.

2 Dip a craft stick into petroleum jelly. Use the stick to smear a thin coat of jelly on a large area of white paper. Do this for each piece of paper.

3 Get an adult to help you attach the strings to the papers and hang the papers in different locations. Leave them for a few hours.

4 **Communicate** Make a table titled What is in the air? On it, record each location.

© Macmillan/McGraw-Hill

5 **Observe** Use a hand lens to observe each paper. Record observations in the table.

Step **2**

Step **4** What Is in the Air?	
near a tree	
near a heating vent	
near a sidewalk	

Draw Conclusions

6 **Interpret Data** What kinds of things are carried in air? What differences, if any, were there among the things found on the papers?

_____

_____

_____

_____

7 **Infer** Might any of the things carried in air make some people sick? Which ones?

_____

_____

Guided Inquiry
What is in water?

Form a Hypothesis

What kinds of things are found in water? Are the same things in different sources of water? Write a hypothesis.

_____

_____

_____

_____

Test Your Hypothesis

Design a plan to test your hypothesis. Use the materials shown. Write the steps you plan to follow.

_____

_____

Draw Conclusions

Did your results support your hypothesis? Why or why not? Share your results with your classmates.

_____

_____

Open Inquiry

What other questions do you have about air pollution or water pollution? Share your questions with other classmates. How might you find the answers to your questions?

Remember to follow the steps of the scientific process.

My question is: _____

_____

How I can test it: _____

_____

_____

_____

My results are: _____

_____

Name _____ Date _____

How can you tell air is around you?

Make a Prediction

Can air keep a paper towel inside a cup from becoming wet?

_____

_____

Materials
• plastic container
• water
• paper towel
• plastic cup

Test Your Prediction

1. Fill a container about two-thirds with water. Stuff a dry paper towel in the bottom of a cup.

2. **Experiment** Hold the cup upside down over the water. Push the cup straight down to the bottom of the container. Do not tilt the cup.

Step 2

3. **Observe** Lift the cup out of the water. Do not tilt it. How does the paper towel feel?

_____

4. **Observe** Repeat step 2. Slowly tilt the cup. Remove it from the water. What do you observe?

_____

Draw Conclusions

5. **Infer** What escaped from the cup in step 4? How did this affect the paper towel?

_____

6 **Infer** How do you know that air is around you?

_____

_____

Explore More

Experiment How else could you show that air is around you? Make a plan to find out.

_____

_____

Open Inquiry

What do you think would happen to the volume of air in a container if the air were heated? Think of your own question about what would happen to the volume of air in the container. Make a plan and carry out an experiment to answer your question.

My question is: _____

_____

How I can test it: _____

_____

My results are: _____

_____

Name _____ Date _____

How does air pressure change?

Make a Prediction

Even when a plastic jug does not contain milk or water, it is not empty. It contains air. What do you think will happen to the air inside the jug when that air is allowed to cool? Make a prediction.

Materials

- plastic milk jug with lid
- funnel
- hot water

_____

_____

Test Your Prediction

1 Ask an adult to use a funnel to carefully add very hot water to the jug until the jug is about half full.
⚠ **Be Careful!** Hot water can cause burns. Then ask the adult to screw the lid back on the jug.

2 Let the jug sit for about one hour so the water inside the jug can cool.

3 **Observe** What happened to the sides of the jug?

_____

Draw Conclusions

4 **Infer** What do you think happened inside the jug to make the walls collapse?

_____

_____

Make a Windsock

1 Bend wire to make a circle. The circle should be about 10 cm across.

2 Cut a sleeve from an old long-sleeved shirt. Staple the sleeve's large opening around the wire. Cut a small opening so you can tie some string to the wire.

3 Tape a small rock across from the string.

4 **Observe** Tie the string to a tree branch. Observe the windsock during the day. Keep a record of what you see.

5 **Infer** What can you tell about the wind from what you observed?

_____

_____

_____

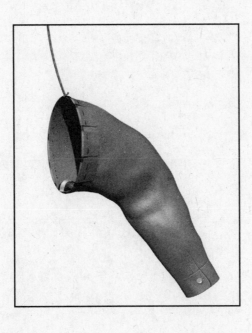

Name _____ Date _____

Interpret Data

Have you ever noticed that some months are warmer or wetter than others? This is generally true from year to year. How did scientists figure this out? One way is to **interpret data** from past years.

Learn It

When you **interpret data**, you use information that has been gathered to answer questions or to solve problems. It is easier to interpret data when it is in a table or a graph. That way you can quickly see differences in the data.

Average Air Temperature (in °C)											
Jan.	Feb.	Mar.	Apr.	May	June	July	Aug.	Sept.	Oct.	Nov.	Dec.
5	7	12	16	21	24	26	26	23	17	18	7

Try It

Scientists collect information about air temperatures in certain places. They use the data to figure out the average air temperature of a certain place for each month of the year. The data on the previous page shows the average monthly air temperatures for the city of Atlanta, Georgia. You can organize and **interpret data** to draw conclusions, too.

First, organize the data by making a bar graph. Follow these steps to make your bar graph.

1 List the months in order along the bottom of the graph. Label the bottom "Month."

2 Write the numbers for the temperatures along the left side of the graph. Write the numbers 0, 2, 4, 6, 8, and so on. End with the number 26. Label this side and write a title for the graph.

3 Draw a bar to match each of the numbers from the data.

4 Now answer these questions: Which months are warmest? Which month is coolest?

_____

_____

Apply It

Now it is your turn to collect and **interpret data**. Measure the air temperature every hour for one school day. Begin at 9:00 A.M. and end at 2:00 P.M. Record your data in a table. Use the table to make a bar graph.

Use your bar graph to interpret your data. When is the air temperature warmest? When is it coolest?

_____

_____

_____

How do raindrops form?

Purpose
Find out how raindrops form in the atmosphere.

Procedure

1 Fill a jar one-fourth full of warm water.

2 Stretch plastic wrap over the top of the jar. Use a rubber band to hold the plastic wrap in place. Place a marble in the center of the plastic wrap.

3 **Make a Model** Place a few ice cubes on the top of the plastic wrap to cool the air above the water. The warm water represents a lake. The air above it represents the atmosphere.

4 **Infer** Observe the bottom of the plastic wrap for several minutes. What forms there? Where did it come from?

_____

_____

_____

Materials

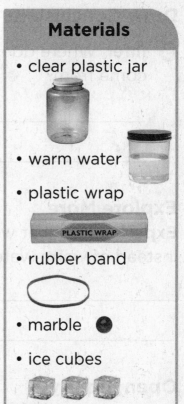

- clear plastic jar
- warm water
- plastic wrap

  PLASTIC WRAP
- rubber band
- marble
- ice cubes

Step **2**

Step **3**

Name _____ Date _____

Draw Conclusions

⑤ **Infer** Where does the water that forms raindrops come from?

_____

_____

Explore More

Experiment What would happen if you used cold water instead of warm water? Try it.

_____

_____

Open Inquiry

What would happen inside the covered jar if it were placed near a sunny window or beneath a lamp? Think of your own question about what would happen inside the jar. Make a plan and carry out an experiment to answer your question.

My question is: _____

_____

How I can test it: _____

_____

My results are: _____

_____

What is the water cycle?

Purpose
In this activity, you will make a model of the water cycle.

Materials

• 2 plastic bottles

• rock

• blue food coloring

• plastic tape

• water

Procedure

1 **Make a Model** Place the rock in the bottom of a plastic bottle. Add water until the rock is about half covered. Add three drops of food coloring to the water and swirl it around.

2 Place a second plastic bottle over the opening of the first bottle. Use tape to seal the two bottles together. Place the bottles in a sunny window.

3 **Observe** After about one hour, observe the bottles. Record your observations.

_____

_____

_____

Draw Conclusions

4 **Infer** Why does this model represent the water cycle?

_____

_____

Name _____ Date _____

Cloud in a Jar

1 **Make a Model** Fill a clear jar half full with warm water. Place a metal tray of ice cubes on top of the jar. Wait for about a minute.

2 **Observe** Darken the room. Shine a flashlight into the jar. What do you see? What is it made of?

_____

_____

3 **Infer** Where do clouds come from?

_____

_____

© Macmillan/McGraw-Hill

How do temperature and precipitation patterns compare?

Purpose

Find out how temperature and precipitation compare between two cities.

Procedure

1 Study the data in these tables. What do they show?

_____

_____

2 **Use Numbers** What are the highest and lowest temperatures for each city? Which city's temperature changes the most during the year? How much precipitation does each city get? Use a calculator.

_____

_____

_____

_____

Average Air Temperature (in °F)											
City	**Jan.**	**Feb.**	**Mar.**	**Apr.**	**May**	**June**	**July**	**Aug.**	**Sept.**	**Oct.**	**Nov.**	**Dec.**
A	21	25	37	49	59	69	73	72	64	53	40	27
B	78	78	78	78	81	82	82	84	84	84	81	80

Average Precipitation (in inches)											
City	**Jan.**	**Feb.**	**Mar.**	**Apr.**	**May**	**June**	**July**	**Aug.**	**Sept.**	**Oct.**	**Nov.**	**Dec.**
A	2	1	3	4	3	4	4	4	4	2	3	2
B	10	10	14	15	10	6	10	10	9	9	15	12

Name _____ Date _____

Draw Conclusions

3 **Interpret Data** How do the temperature and precipitation patterns for these cities compare?

_____

_____

_____

4 **Infer** Which city is better for growing orchids? Why?

_____

_____

Explore More

Interpret Data What are the monthly air temperatures and precipitation like where you live? How could you find out?

_____

_____

Open Inquiry

In your town, is it warmer during some seasons of the year than in others? Think of your own question about seasons and temperatures in your town. Make a plan and carry out an experiment to answer your question.

My question is: _____

How I can test it: _____

_____

My results are: _____

© Macmillan/McGraw-Hill

How do temperature and precipitation change by season?

Materials

• graph paper

Purpose

The calendar can be divided into four seasons: winter, spring, summer, and fall. Each season is about three months long. Winter is December, January, and February. Spring is March, April, and May. Summer is June, July, and August. Fall is September, October, and November. Find out how temperature and precipitation compare for each season between two cities.

Procedure

① Examine the tables on page 125. They show average air temperatures and average precipitation for each month of the year for two different cities.

② **Use Numbers** Look at the data in the tables. What is the warmest season for City A and City B? What is the coldest season for each city?

_____

_____

Draw Conclusions

③ **Infer** Which city is located in a warm tropical area?

_____

_____

Name _____ Date _____

Compare Climates

1 **Make a Model** Label one sheet of paper *City A* and another *City B*. Use a flashlight to model the Sun. Hold it about 6 centimeters above the paper for *City A*. Shine it straight down. Have a partner use a pencil to trace along the edge of the light.

2 **Make a Model** Repeat step 1 for *City B*. This time, tilt the flashlight as you shine it on the paper.

3 **Interpret Data** Over which city is the shape of light larger? Over which city is the Sun's energy more spread out?

_____

_____

4 **Infer** Which city would have a colder climate?

_____

© Macmillan/McGraw-Hill

How do shadows change throughout the day?

Form a Hypothesis

How does the Sun's location affect the length of shadows cast on the ground? How does it affect their position? Write a hypothesis.

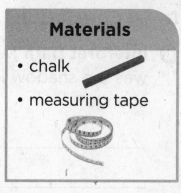

Materials

- chalk
- measuring tape

_____

_____

_____

Test Your Hypothesis

1 Work in pairs outside on a sunny morning. Use chalk to mark an *X* on the pavement.

2 Have your partner stand on the X. Trace your partner's shadow.

3 **Measure** Use a measuring tape to find the length of the shadow. Observe the location of the Sun and the shadow. Record the data in a table.
⚠ **Be Careful.** Never look directly at the Sun.

4 **Observe** Repeat steps 2 and 3 at midday and in the afternoon. How did the Sun's location and the shadow change?

_____

_____

Draw Conclusions

5 **Interpret Data** When was the shadow shortest? When was the shadow longest? How did the Sun's position change?

_____

_____

6 **Infer** What caused the shadow to change position and length?

_____

_____

Explore More

Experiment How can you find out in which month the Sun's position appears highest in the sky? Lowest?

_____

Open Inquiry

How can you make shadows on a cloudy day without the Sun? Think of your own question about how to make your own shadows. Then, design and carry out an experiment to answer your question.

My question is: _____

_____

How I can test it: _____

My results are: _____

_____

How can you model shadows with a lamp?

Form a Hypothesis

Form a hypothesis about how your shadow will change as you move around a table lamp. The table lamp is a model of the Sun. Start your hypothesis with "If I move around the lamp, then . . ."

Materials
• table lamp
• chalk or masking tape
• measuring tape

_____

_____

Test Your Hypothesis

1. **Observe** Place the table lamp on the floor in the center of the room. Clear an area around the lamp. Make a chalk mark on the floor. (If you cannot make a mark with chalk, use a piece of masking tape to make a mark.)

2. **Measure** Have your partner stand on the chalk mark. With the measuring tape, measure the length of your partner's shadow. Continue moving and measuring until you have four measurements. How does the shadow change as your partner moves?

_____

_____

Draw Conclusions

3. **Infer** How does moving the lamp model the movement of the Sun?

_____

_____

Name _____ Date _____

Model Earth's Rotation

1 **Make a Model** Carefully push a pencil through a foam ball. The ball represents Earth. The pencil represents Earth's axis. Press a paper clip into the side of the ball. The paper clip represents you.

2 **Observe** Shine a light on the paper clip. The light represents the Sun. On the model, where is it day?

_____

_____

3 **Observe** Rotate the pencil so the paper clip faces away from the light. Where is it night?

_____

_____

4 **Interpret Data** What does this model help to show?

_____

_____

© Macmillan/McGraw-Hill

How does the Moon's shape seem to change?

Make a Prediction

How will the Moon seem to change during one school week? Write a prediction.

_____

_____

Test Your Prediction

1 **Observe** Work with an adult. Look at the Moon each night for one school week. You can also use the Internet to find out what the Moon looks like each night.

2 **Communicate** Record what the Moon looks like each night. Also record where the Moon is in the sky and the time that you observed it.

_____

_____

Draw Conclusions

3 **Interpret Data** How did the Moon's shape seem to change each night? How does its location seem to change?

_____

_____

4 **Predict** What will the Moon's shape be like for the next few nights?

_____

_____

Explore More

Experiment How does the Moon's shape and position in the sky change over several months? Make a plan to find out.

_____

_____

_____

Open Inquiry

How long does it take for the Moon to go from a full Moon to another full Moon? Think of your own question about this topic, then design an experiment to answer your question.

My question is: _____

_____

How I can test it: _____

_____

My results are: _____

_____

© Macmillan/McGraw-Hill

How does the Moon change in a month?

Purpose

Track changes in the Moon's appearance for a month.

Materials

- notebook
- pencil
- reference materials such as books or the Internet

Procedure

1 **Observe** Each night for a month, make a drawing of the Moon in a notebook. Be sure to label each drawing with the date.

2 **Infer** How long did it take for the full Moon to change to a non-visible Moon (new Moon)?

_____

_____

3 **Classify** Use reference books or the Internet to classify the different changes of the Moon. Label the phases of the Moon in your notebook. What are the phases called?

_____

_____

Draw Conclusions

4 **Communicate** How long does it take for the Moon to change from a full Moon to the next full Moon?

_____

Name _____ Date _____

Make a Flip Book

1 Draw what each Moon phase looks like on the right side of an index card. Label the phase. Do this for each of the eight phases.

2 Put the cards in order. Staple them together on the left side.

3 Flip the pages with your thumb. How does the Moon's shape seem to change?

_____

_____

_____

© Macmillan/McGraw-Hill

Structured Inquiry

Why does the Moon's shape appear to change?

Materials
- lamp
- ball

Form a Hypothesis

The Moon completes one orbit around Earth about every $29\frac{1}{2}$ days. How does the Moon's position in space affect how we see the Moon? When do we see a full Moon? When do we see a last-quarter Moon? Write a hypothesis.

_____

_____

_____

Test Your Hypothesis

1. **Make a Model** Hold the ball and stretch out your arm in front of you. Position your arm so that the ball is a little higher than your head. The ball is the Moon, and your head is Earth.

2. **Observe** Your teacher will turn on a lamp. The lamp represents the Sun. Turn your back to the light so that the light shines on the ball. What part is lit up?

   _____

   _____

Step 2

© Macmillan/McGraw-Hill

3 **Experiment** Turn in place while holding the ball. Keep the ball in front of you. Notice the change in light and shadow on the ball.

4 **Communicate** Record your observations by drawing or describing what you saw as you turned.

_____

_____

5 Repeat the experiment several times. Are your observations the same with each trial?

_____

_____

Draw Conclusions

6 **Observe** Where is the ball when it is lit up like a full Moon? Where is the ball when it looks like a last-quarter Moon?

_____

_____

_____

7 **Infer** Why does the Moon's shape appear to change? How do you know?

_____

_____

_____

© Macmillan/McGraw-Hill

Guided Inquiry

How does the Moon's position change?

Form a Hypothesis

How does the Moon's position change in the night sky over the course of one month? Write a hypothesis.

_____

_____

_____

Test Your Hypothesis

Design a plan to test your hypothesis. Remember your plan should test only one variable—the Moon's position in the sky. Decide on the materials you will need. Then write the steps you plan to follow.

_____

_____

_____

Draw Conclusions

Did your results support your hypothesis? Why or why not? Share your results with your classmates.

_____

_____

_____

© Macmillan/McGraw-Hill

Name _____ Date _____

Open Inquiry

What other questions do you have about the Moon? Talk with your classmates about questions you have. How might you find the answers to your questions?

My question is: _____

_____

How I can test it: _____

_____

_____

My results are: _____

_____

_____

_____

How do the planets move through space?

© Macmillan/McGraw-Hill

Materials

- masking tape

- 8 planet cards and 1 Sun card

Purpose

Model how the positions of the planets change.

Procedure

1 Work in a large room or other area. Put a chair in the center of the room. Tape the card labeled Sun to the chair. Tape a line from the chair to a wall.

2 Form two groups. Each student in the first group will take a card and line up along the tape in order: Mercury, Venus, Earth, Mars, Jupiter, Saturn, Uranus, and Neptune.

3 **Make a Model** Students in the first group model how the planets move by walking in a circle around the Sun. Take steps of the same size. Count steps together. Stop counting when you reach the tape again. Students in the second group repeat steps 2–3.

4 **Use Numbers** Do all the "planets" complete one trip around the Sun in the same number of steps?

_____

_____

Draw Conclusions

5 What was different about the orbits?

_____

_____

6 Infer How do planets move through space?

_____

_____

Explore More

Predict What planets can you see in the night sky where you live? Make a plan to find out.

_____

_____

Open Inquiry

What would happen if you took into account the speeds of the planets' orbits as you conducted this activity? Think of your own question about this topic. Then, design and carry out an experiment to answer your question.

My question is: _____

_____

How I can test it: _____

_____

_____

My results are: _____

_____

_____

_____

How fast do the planets move?

Materials

- reference books
- the Internet

Purpose

Conduct research to determine and compare the average orbital speeds of the planets in the solar system.

Procedure

1 **Use Numbers** Research the average orbital speeds of the planets in our solar system. In what units are the speeds listed?

_____

2 **Measure** What are the average orbital speeds of the planets? List them in the table.

Planet	Speed
Mercury	
Venus	
Earth	
Mars	
Jupiter	
Saturn	
Uranus	
Neptune	

3 **Interpret Data** Which planet has the slowest average orbital speed?

_____

4 **Interpret Data** Which planet has the fastest average orbital speed?

_____

© Macmillan/McGraw-Hill

Quick Lab

Sizing Up Planets

1 **Measure** Work with a partner. Hold a marble about 30 centimeters away from you.

2 **Measure** Have a partner hold a tennis ball about 5 meters away from you.

3 **Observe** Which object seems larger? Why? Which object really is larger?

_____

_____

_____

4 **Infer** How can larger planets look smaller to us than smaller planets?

_____

_____

_____

© Macmillan/McGraw-Hill

Observe

You know that Earth is only one of the planets in our solar system. How do scientists learn about other planets? How do they learn about other objects, such as comets, that orbit the Sun? They **observe** the Sun, planets, moons, and other objects in our solar system to learn more about them.

Learn It

When you **observe**, you use one or more of your senses to learn about an object or event. Remember, your senses are sight, hearing, smell, taste, and touch. Scientists often use tools, such as binoculars, microscopes, and telescopes, to make their observations. They use their observations to draw conclusions about objects and events.

⚠️ **Be Careful.** It can be dangerous to observe things by tasting. You should not taste things in school unless your teacher tells you it is safe.

© Macmillan/McGraw-Hill

Name _____ Date _____

Try It

You can **observe** things, too. Look at the detail of this comet in your textbook. Observe its color and shape. Look for unique features that help you identify what it is. What detail helps you know this is a comet and not a planet?

_____

_____

Apply It

The photos on the next page show details of planets and other objects in our solar system. **Observe** each photo carefully. Use your observations to identify what each object is. Which details helped you to identify the objects?

_____

_____

_____

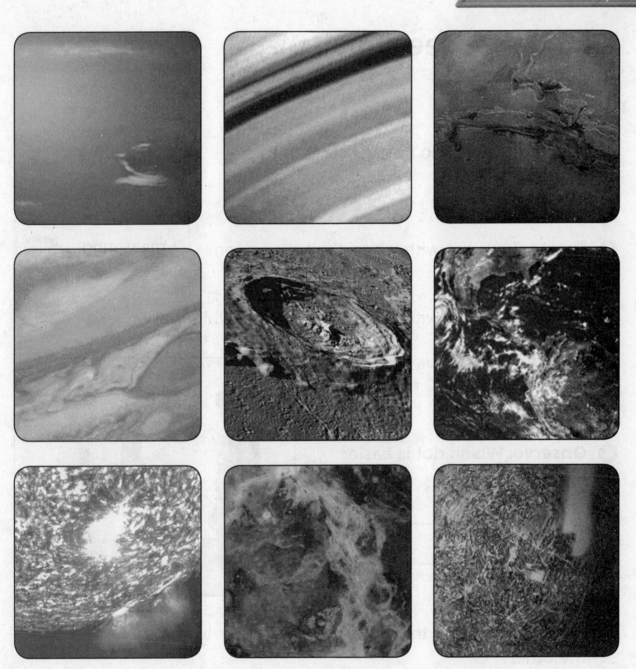

Name _____ Date _____

Why do we see stars at night?

Materials

- white chalk or white crayon

- black paper

- white paper

- measuring tape

Purpose

Understand why we do not see stars in the daytime sky.

Procedure

1 Draw a 3 cm dot with white chalk or crayon on black paper.

2 Draw a 3 cm dot with chalk or crayon on white paper.

3 **Measure** Have your partner hold both papers 2 m away from you.

4 **Observe** Which dot is easier to see?

_____

_____

Step 3

Draw Conclusions

5 **Infer** Why do you think one dot was easier to see than the other?

_____

_____

6 **Infer** Suppose the dots on the papers were stars. Why do you think you see stars only at night?

_____

_____

_____

Explore More

Experiment Design an experiment to show how we sometimes see the Moon during the day.

_____

_____

Open Inquiry

Is it any easier to see an entire group of stars than a single star during the day? Think of your own question about this topic. Then, design and carry out an experiment to answer your question.

My question is: _____

_____

How I can test it: _____

_____

My results are: _____

_____

_____

Why are some stars brighter than others?

Materials

• reference books or the Internet

Form a Hypothesis

Form a hypothesis about why some stars in the night sky are brighter than others. Start your hypothesis, "If some stars in the night sky are brighter than others, then . . ."

_____

_____

_____

Test Your Hypothesis

1 **Observe** Use reference books or the Internet to find out why some stars look brighter from Earth than other stars.

2 **Interpret Data** Imagine that two stars are the same size and brightness, but they do not look like they have the same brightness from Earth. Why might they look different?

_____

_____

3 **Interpret Data** Imagine that two stars are the same distance from Earth, but one looks brighter than the other. Why might one look brighter?

_____

_____

Make a Constellation

1 Find out about a constellation. Use chalk to draw it on a piece of black paper. Use a pencil point to poke holes in the pattern. ⚠ **Be Careful.**

2 **Observe** Trade papers with a partner. In a darkened room, hold your partner's paper out at arm's length toward a lamp or flashlight. What does the star pattern look like?

_____

_____

_____

3 **Communicate** Tell your partner the name of your constellation.

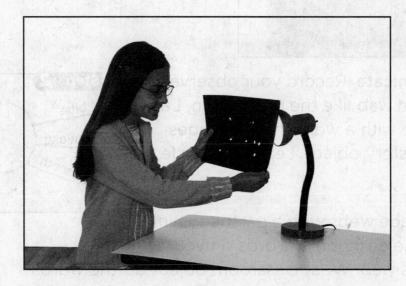

© Macmillan/McGraw-Hill

Name _____ Date _____

How do you describe objects?

Materials

Purpose

Explore ways to describe objects.

Procedure

1 **Observe** Select a "mystery object" in your classroom. Observe the object. What color is it? How does it feel? What is the object's shape and size?

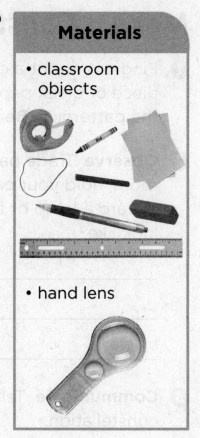

- classroom objects

- hand lens

_____

_____

2 **Communicate** Record your observations in a word web like the one shown. Label each line with a word that describes your mystery object. Leave the circle blank.

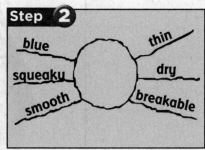

blue   thin

squeaky   dry

smooth   breakable

3 **Infer** Trade webs with a partner. Think about the descriptive words on your partner's web. What classroom object do the words describe? Label the circle with the name of your partner's mystery object.

Draw Conclusions

4 Were you able to guess your partner's mystery object? Was your classmate able to guess your mystery object?

_____

_____

Explore More

Experiment How might your web be different if you were blindfolded and could only touch the mystery object? Try it to find out.

_____

_____

Open Inquiry

How might your descriptions change if the object were in a box and you could neither see it nor feel it? Think of a question about a hidden object. Make a plan and carry out an experiment to answer your question.

My question is: _____

_____

How I can test it: _____

_____

_____

My results are: _____

_____

What are the properties of an object?

Purpose

In this activity, you will describe the properties of an object so that someone else can infer what it is.

Procedure

1 Work with a partner. You and your partner will take turns secretly choosing an object in the classroom.

2 Choose an object and have your partner ask up to 5 questions about its properties. Each question must only have a "yes" or "no" answer.

3 **Record Data** Use the table below to help you.

Guess From Properties		
Property	**Question**	**Yes or No**

4 **Communicate** What object did your partner choose?

_____

5 **Interpret Data** How did asking questions about the properties of the object help you to identify it?

_____

Classify Matter

① Look at ten objects.

② **Communicate** List the properties of each object in a table like the one shown below.

③ **Classify** Sort the objects into groups that have similar properties. Give each group a name that describes how its items are alike.

④ **Interpret Data** Did some of the objects in one group have the same properties as objects in another group? How did you decide how to classify each object?

_____

_____

⑤ **Communicate** Is there more than one way to classify these objects? Explain your answer.

_____

_____

Object	Properties

Name _____ Date _____

How can you measure length?

Make a Prediction

How wide is your classroom? Write a prediction.

_____

_____

Test Your Prediction

1 **Measure** Work with a partner. Stand with your back against one wall. Slowly walk across the room, placing one foot in front of the other. The heel of your front foot should meet the toe of your back foot. Your partner will count the number of steps it takes to cross the room.

2 Trade roles with your partner and repeat step 1.

3 **Communicate** Compare your data with your class's data. Make a table listing the data for the entire class.

Draw Conclusions

4 **Interpret Data** What is the highest measurement? What is the lowest measurement? Did anyone get the same measurement?

_____

_____

_____

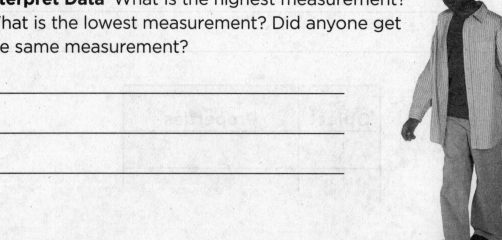

5 **Infer** Why were there different measurements? Why do we not use our feet to measure length?

_____

_____

_____

Explore More

Measure Scientists use the metric system to measure matter. Predict how wide your classroom is in meters and centimeters. Then use a metric ruler to measure the width of your classroom. How do your measurements compare with your predictions?

_____

_____

Open Inquiry

Would your results would be easier to compare if everyone used unsharpened pencils instead of their feet to measure distance? Make a plan and carry out an experiment to answer your question.

My question is: _____

_____

How I can test it: _____

_____

_____

My results are: _____

© Macmillan/McGraw-Hill

Name _____ Date _____

How is length measured?

Purpose

Based on a given measurement, infer what tool
was used.

Procedure

1 **Measure** While your partner closes his or
her eyes, use a pencil, ruler, or paper clip to measure the
length of your desk. How many units did it measure?

_____

2 **Infer** Have your partner open his or her eyes and look at
your measurement. Have your partner infer which tool
you used.

3 **Measure** Switch places with your partner. How many
units did your partner use to measure the desk?

_____

4 **Infer** What tool do you think your partner used?

_____

5 **Measure** Choose three other objects to use as tools.
Repeat this activity. Write your measurements and
guesses of the tools on the lines below.

Your measurement: _____

Your partner's guess: _____

Your partner's measurement: _____

Your guess: _____

© Macmillan/McGraw-Hill

Measure Mass and Volume

1 **Predict** Look at a toy car, golf ball, and marble. Predict which object has the most mass. Which has the greatest volume?

_____

_____

2 **Measure** Find the mass of each object. List the objects from most mass to least mass.

_____

3 **Measure** Fill a measuring cup with 250 mL of water. Add one object at a time to the measuring cup. Record the water level for each object.

_____

_____

4 **Interpret Data** List the objects from greatest to least volume.

_____

5 **Interpret Data** Which object has the most mass? Which object has the greatest volume? How did the results compare with your prediction?

_____

_____

Measure

You have learned that matter is anything that takes up space and has mass. Water is matter that is important to life on Earth. It is found on Earth as solid ice and liquid water. It is even found in the air. What happens to water's mass as it changes from a chunk of solid ice to liquid water? Scientists **measure** things to answer questions like this.

Learn It

When you **measure,** you find such things as the mass, volume, length, or temperature of an object. You can also measure distances and time. Scientists use many tools to measure things. Some of these tools are shown on this page. Scientists use measurements to describe and compare objects or events.

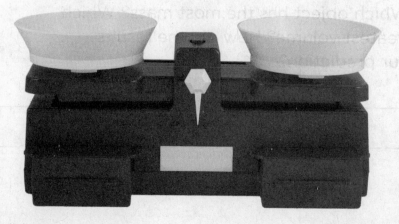

© Macmillan/McGraw-Hill

Try It

You know that scientists **measure** things to answer questions. You can measure, too. Answer this question. Do ice cubes have the same mass after they melt?

1 To start, place several ice cubes in a cup. Then, cover the cup with plastic wrap so the water stays inside the cup.

2 Measure mass by placing the cup on one end of a pan balance. Add masses to the other side of the pan balance until both sides are level. Record the mass on a chart.

3 Measure the mass every $\frac{1}{2}$ hour until the ice is completely melted.

4 Now use your measurements to answer the question. Do ice cubes have the same mass after they melt?

_____

_____

Time	Mass

Apply It

Now **measure** to answer this question: Does ice cream have the same mass after it melts? How do you know?

_____

_____

How are solids different from liquids?

Make a Prediction

How do you know if something is solid? How do you know when something is a liquid?

_____

_____

_____

Test Your Prediction

1 **Observe** Touch the block. Does it feel more like a solid or more like a liquid? Why?

_____

_____

2 **Experiment** Put the block into the beaker. Record your observations.

3 **Experiment** Use the spoon to stir the block. What happens? Record your observations. Empty the beaker.

_____

_____

4 Repeat steps 1–3. Instead of the block, use the water, salt, hand soap, and clay. Test each object one at a time.

Draw Conclusions

5 Which objects did not change shape? Which objects were easy to stir?

_____

_____

Materials

- block

- plastic spoon

- water

- beaker

- hand soap

- clay

- safety goggles

6 Classify Which objects are solids? Which are liquids?

_____

7 Explain how solids are different from liquids.

_____

Explore More

Experiment What would happen if you put each object in the freezer? What would happen if you put each object in a warm place? Form a hypothesis and test it.

_____

_____

Open Inquiry

Extend the activity to include any changes in volume of the items tested. Think of a question about volume. Make a plan and carry out an experiment to answer your question.

My question is: _____

_____

How I can test it: _____

_____

My results are: _____

Solid or liquid?

Purpose
Decide whether a material is a solid or a liquid.

Purpose

1 **Experiment** Put some cornstarch in a container. Add water. Use your hands to mix the ingredients well.

2 **Observe** Look at what you mixed. Feel it. List its properties.

_____

_____

3 Wash your hands.

4 **Interpret Data** Is the mixture a solid or a liquid, or does it have properties of both states?

_____

5 **Illustrate** In the space below, draw a picture of the container and the product.

© Macmillan/McGraw-Hill

Name _____ Date _____

Compare Solids, Liquids, and Gases

Materials

• 3 plastic bags

• water

• small rock

1 Blow into an empty bag. Then quickly seal the bag.

2 Fill a second bag with water and seal this bag. Put a rock in a third bag and seal it.

3 **Observe** Each bag contains matter in a different state. How does each bag look and feel? Record your observations.

_____

_____

_____

4 **Observe** Open each bag. What happens?
△ **Be Careful!** Hold the bag filled with water over a container.

_____

_____

_____

5 **Communicate** Describe the properties of a solid, a liquid, and a gas. Tell how these three states of matter are different from one another.

_____

_____

_____

© Macmillan/McGraw-Hill

What happens when ice is heated?

Materials

- spoon
- thermometer
- plastic cup of ice cubes

Make a Prediction

How does ice change as it is heated? Write a prediction.

_____

Test Your Prediction

1 **Measure** Place a thermometer in the cup of ice. Measure the temperature of the ice. Record the temperature in a table like the one shown.

2 Place the cup in a warm place, such as on a sunny windowsill.

3 **Measure** Stir the ice and measure its temperature every 10 minutes for the next hour. Record the temperature in the table.

4 Describe how the ice changes.

Step **1**	
Time	**Temperature**

Step **3**

_____

Draw Conclusions

5 **Communicate** How did the ice change as it was heated? Was your prediction correct?

_____

6 **Infer** What happened to the temperature of the water as the ice melted? Why do you think this happened?

_____

_____

Explore More

Predict What will happen to the water as it continues to sit in the Sun after the ice has melted? Test your prediction and find out.

_____

_____

_____

Open Inquiry

How would the results be different if you had used warm water instead of ice? Think of your own question about how water temperature would affect the results. Make a plan and carry out an experiment to answer your question.

My question is: _____

_____

How I can test it: _____

_____

My results are: _____

_____

How does energy affect state?

Make a Prediction

Predict What change of state happens when you place a drop of water on your forearm and spread it around?

_____

_____

Test Your Prediction

1 **Experiment** Place a drop of water on your forearm. Your forearm is the inside of your arm between the wrist and the elbow. Spread the water around.

2 **Observe** Does the water feel warm or cool on your arm?

_____

3 **Observe** What happens to the water after a few minutes?

_____

Draw Conclusions

4 Which have more energy, the particles of water that entered the air, or the particles of water that remain on your arm?

_____

Name _____ Date _____

Condense Water Vapor

1 **Observe** Feel an empty plastic cup. Does it feel wet or dry? Does it feel hot or cold? Record your observations.

_____

_____

2 Fill your cup with ice cubes. Next add cold water to the cup.

3 **Observe** Feel your cup again. Does the cup feel wet or dry? Does the cup feel hot or cold? Record your observations.

_____

4 **Observe** Look at your cup after five minutes. What do you notice about the outside of the cup? Is it wet or dry?

_____

_____

5 **Infer** Where did the water on the cup come from?

_____

Predict

You just learned about how liquids change to solids. Which do you think freezes faster, salt water or fresh water? To find answers to questions like this, scientists **predict** what they think will happen. Next, they experiment to find out what does happen. Then, they compare their results with their prediction.

Learn It

When you **predict,** you state the possible results of an event or experiment. It is important to record your prediction before you do an experiment. Next, you record your observations as you experiment and record the final results. Then you have enough data to figure out if your prediction was correct.

Ice floats on the salt water of Shoup Bay in Alaska.

Try It

Which do you think freezes faster, salt water or fresh water? **Predict** what will happen. Write your prediction on a chart like the one shown. Then do an experiment to test your prediction.

1 Pour 125 mL of water into a plastic cup. Label this cup *Fresh Water.*

2 Pour 125 mL of water into another plastic cup. Add 1 tablespoon of salt and stir with a spoon. Label this cup *Salt Water.*

3 Place both cups into the freezer. Check them every 15 minutes. Draw or write your observations on a separate piece of paper.

▶ Which freezes faster, fresh water or salt water?

▶ Was your prediction correct?

_____

Materials
• measuring cup
• water
• two plastic cups
• salt
• measuring spoon

Which Freezes Faster?	
My Predictions	
Observations of Fresh Water	
Observations of Salt Water	
Results	

Apply It

Now that you have learned to think like a scientist, make another prediction. Do you **predict** that salt water or fresh water will evaporate faster? Plan an experiment to find out if your prediction is correct.

_____

_____

_____

_____

_____

_____

© Macmillan/McGraw-Hill

Name _____ Date _____

How can you change matter?

Purpose

Find out some ways you can change matter.

Procedure

1 Make a table like the one shown below.

2 **Observe** Look at each object. What properties does each object have? How can you change each object?

_____

_____

3 **Experiment** Change each object. What properties does each have now? Record the property that has changed. ⚠ **Be Careful!** Handle scissors carefully.

_____

_____

_____

Materials

- paper

- clay

- ice cubes

- scissors

Step **3**

Object	Change	Properties changed
Paper		
Clay		
Ice Cubes		

© Macmillan/McGraw-Hill

Draw Conclusions

4 How are the objects different after you made the changes?

_____

5 **Infer** Do you think you changed the kind of matter making up the object? Explain.

_____

_____

Explore More

Experiment What would happen if you add a spoon of salt to a cup of water? How does the salt and water change? How can you remove the salt from the water?

_____

_____

Open Inquiry

Do you think that a change of state changes the identity of matter? Think of your own question about changes to matter, and then make a plan and carry out an experiment to answer your question.

My question is: _____

_____

How I can test it: _____

_____

My results are: _____

_____

How does changing state change matter?

Materials
- narrow-mouth jar
- ice cube
- hot water

Purpose

Observe how a liquid changes to a gas and back to a liquid.

Procedure

① **Experiment** Set the jar on your desk or lab table. Have your teacher pour hot water into the jar until it is about half full. ⚠ **Be Careful!** Do not touch the outside of the jar.

② Place an ice cube on the top of the jar. It should rest on the mouth of the jar but not fall into the jar.

③ **Observe** What do you observe happening on the sides of the jar?

_____

④ **Infer** Why did you place the ice cube on the top of the jar?

_____

_____

⑤ **Conclude** How do you know that the changes you observed are physical changes?

_____

_____

Separating Mixtures

1 Mix some sand, marbles, and paper clips together in a bowl.

2 **Experiment** Design an experiment to separate this mixture.

_____

_____

3 **Observe** Were you able to completely separate the mixture? How do you know?

_____

_____

4 **Experiment** How could you separate a mixture of sugar and water?

_____

_____

Name _____ Date _____

How can matter change?

Make a Prediction

How do flour and baking soda change when each is mixed with vinegar? Write a prediction.

_____

_____

_____

- vinegar
- flour
- baking soda
- 2 plastic bottles
- 2 balloons
- goggles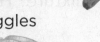
- measuring cup and spoons
- funnel

Test Your Prediction

1 **Observe** List the properties of the vinegar, flour, and baking soda.

_____

_____

2 **Measure** Use a funnel to put 2 tablespoons of flour in one balloon. Add 50 mL of vinegar to a plastic bottle.

3 **Experiment** Carefully, put the balloon over the bottle's opening without letting any flour fall into the bottle. After you attach the balloon, raise it up so the flour goes into the bottle. Record your observations. Repeat with a second balloon and baking soda.

Step **3**

_____

_____

_____

Draw Conclusions

5 Did your results match your prediction? Explain your answer.

_____

6 **Infer** What do you think caused the differences in the balloons?

_____

_____

Explore More

Experiment What might happen to the balloon if you add two tablespoons of baking powder and 50 mL of water.

_____

Open Inquiry

Why are water and baking powder mixed into many batters and doughs before they are baked? What would happen if the baking powder were left out of the batter? Think of your own question about the presence of baking powder in dough. Make a plan and carry out an experiment to answer your question.

My question is: _____

_____

How I can test it: _____

_____

_____

My results are: _____

_____

© Macmillan/McGraw-Hill

Name _____ Date _____

What is evidence of a chemical change?

Make a Prediction

What happens when a piece of paper burns? Predict what type of change will happen to a piece of paper when it burns.

_____

_____

Test Your Prediction

1 **Observe** Closely look at the piece of paper your teacher is showing the class. What are some properties of the paper?

_____

_____

2 **Observe** Closely watch as your teacher burns the paper. What evidence of change do you observe?

_____

_____

Draw Conclusions

3 **Analyze** What type of change happened when the paper burned?

_____

_____

4 **Communicate** Why do you think this kind of change occurred?

_____

_____

_____

© Macmillan/McGraw-Hill

A Chemical Change

1 **Observe** Look closely at some pennies. Make a list of their properties.

2 Place 1 teaspoon of salt in a bowl. Add 150 mL of vinegar. Stir until the salt dissolves.

3 **Experiment** Dip a penny halfway into the liquid. Slowly count to 20 as you hold the coin there. Then remove the penny. Compare the half you held with the half that was in the liquid.

_____

_____

_____

4 **Infer** What caused the change in appearance?

_____

_____

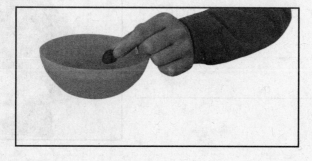

© Macmillan/McGraw-Hill

Structured Inquiry

How can physical and chemical changes affect matter?

Materials

- chalk
- hand lens
- black construction paper
- vinegar
- dropper

Form a Hypothesis

How will breaking chalk change the chalk? How will adding vinegar to the chalk change it? Write a hypothesis.

_____

_____

Test Your Hypothesis

1 **Observe** Break a piece of chalk in half. Use a hand lens to look at the broken end of the chalk. Record your observations. Is this a chemical or physical change?

_____

2 **Experiment** Rub one of the chalk pieces on a piece of black paper. Using the hand lens, look at the chalk on the paper. Record your observations. Is this a chemical or physical change?

_____

_____

Step **2**

© Macmillan/McGraw-Hill

3 **Experiment** Use a dropper to add one drop of vinegar to the chalk on the black paper. Record your observations. Is this a chemical or physical change?

Step **3**

_____

_____

_____

Draw Conclusions

4 **Interpret Data** What did you observe? Which changes were physical changes? Was there a chemical change?

_____

_____

5 **Infer** Describe what happened to the chalk when you added vinegar. What caused this to happen?

_____

_____

6 **Communicate** Use your observations to write your own definitions of chemical and physical change.

_____

_____

_____

_____

Name _____ Date _____

Guided Inquiry

What are the signs of a chemical change?

Materials

• plastic cups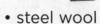

• spoon

• milk

• steel wool

• vinegar

• baking soda

Form a Hypothesis

How can you tell a chemical change has happened? Write a hypothesis.

_____

_____

_____

Test Your Hypothesis

Design an experiment to investigate chemical changes. Use the materials shown. Write the steps you plan to follow. Record your results and observations.

_____

_____

_____

_____

Draw Conclusions

What changes did you observe? Did your experiment support your hypothesis? Why or why not?

_____

_____

_____

_____

Be a Scientist

Open Inquiry

What else would you like to know about physical and chemical changes? Come up with a question to investigate. For example, how does iron rust? Design an experiment to answer your question.

Remember to follow the steps of the scientific process.

My question is: _____

_____

How I can test it: _____

_____

My results are: _____

_____

Name _____ Date _____

How can you describe an object's position?

Materials
- notebook
- two sets of 10 colored blocks

Purpose
Find out ways to describe a block's position.

Procedure

1. Sit opposite a partner at a table. Prop up a notebook between the two of you.

2. One partner, "the builder," uses the blocks to make a building. Make sure the other partner, "the copier," cannot see the building.

Step 2

3. **Communicate** The builder tells the copier how to make the same building. Make a list of the words you use.

4. **Observe** Remove the notebook. Are the buildings the same? Switch roles and try the activity again.

Step 3

_____

_____

Draw Conclusions

5 What words did you use to describe your building?

_____

_____

6 **Infer** Could you describe the position of each block without comparing it to other blocks around it?

_____

_____

Explore More
Communicate How could you direct someone from your home to your school?

_____

_____

Open Inquiry

Think of questions you could ask to find an object in your classroom. Then have a partner choose an object. Ask yes-or-no questions about its position until you find the item.

My question is: _____

How I can test it: _____

_____

My results are: _____

_____

Name _____ Date _____

Where is the treasure?

Purpose

Locate an object by finding its position compared to known points.

Procedure

1 Starting at the lower left corner of a piece of graph paper, label the vertical lines at the bottom with letters. Start with A and continue alphabetically toward the lower right corner of the page.

2 Starting at the lower left corner, label the horizontal lines on the left side of the paper with numbers. Start with 0 and continue numbering up the page toward the top left corner.

3 Somewhere on your graph paper, draw a rectangle that is four squares by two squares. This is your treasure chest! Do not let other students see where your treasure chest is.

4 Work with a partner. Hide your papers from each other, and take turns guessing where your partner's treasure chest is by choosing letter-and-number pairs.

5 **Infer** Could you have located your partner's treasure chest if you had not labeled the graph paper? Explain.

_____

_____

_____

Measure Speed

1 Set up a racetrack as shown below.

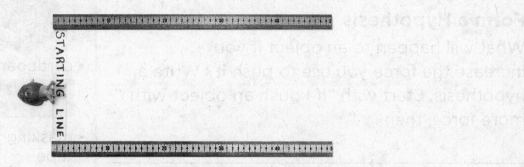

2 **Measure** Wind up a wind-up toy. Place it at the starting line and let it go. Have a partner use a stopwatch to time the toy's trip. Measure how far the toy travels. Record your measurements.

3 **Communicate** Make a drawing on a separate piece of paper to show how the toy moved.

4 **Use Numbers** How far did the toy travel? How fast did it travel? What two measurements do you need to find the toy's speed?

_____

_____

_____

Name _____ Date _____

How can pushes affect the way objects move?

© Macmillan/McGraw-Hill

Materials

- six books
- cardboard
- masking tape
- toy car
- tennis ball
- ruler

Form a Hypothesis

What will happen to an object if you increase the force you use to push it? Write a hypothesis. Start with "If I push an object with more force, then . . ."

_____

_____

_____

Test Your Hypothesis

① Stack three books on the floor. Then lean a piece of cardboard against the top book to make a ramp. Tape down the edge along the floor.

② **Observe** Place a toy car at the bottom of the ramp. Hold the tennis ball at the top of the ramp. Then let the ball go so that it pushes the toy car. What happens?

③ **Measure** Find out how far the car travels.

_____

_____

④ **Use Variables** Add 3 more books to the stack. Repeat steps 2–3. What happens?

_____

_____

Draw Conclusions

⑤ Infer What caused the car to move?

_____

⑥ Interpret Data When did the car travel farther?

_____

⑦ Infer How does the amount of force you use to push an object affect how far the object travels?

_____

_____

Explore More

Experiment What would happen if you added a weight to the toy car and repeated the activity?

_____

Open Inquiry

How does the surface over which an object moves affect a push? Think of a question about how a push is affected by the surface an object moves over. Make a plan and carry out an experiment to answer your question.

My question is: _____

_____

How I can test it: _____

_____

My results are: _____

How does mass affect a push?

Purpose
Show how mass is related to the amount of force on an object.

Procedure

1 Set up a ramp by leaning a piece of folded or grooved cardboard against a stack of three thick books. Tape the other end of the cardboard to your desk. Place a toy truck at the end of the ramp.

2 **Experiment** Roll a tennis ball down the ramp so that it hits the back of the truck. Measure how far the truck moved. Record this measurement.

3 **Experiment** Place the truck back at the bottom of the ramp. This time, roll a baseball down the ramp so that it hits the back of the truck. How far does the truck move? Record this measurement.

4 **Compare** Which ball has the greater mass?

_____

5 **Draw Conclusions** From your data and observations, which will give a greater push, an object with less mass or one with greater mass?

_____

Name _____   Date _____

Observe Gravity

1 **Predict** Does gravity act the same on all objects?
Would it act the same on two plastic bottles that have
the same volume but different mass?

_____

_____

_____

2 Hold an empty plastic bottle in one hand. Hold an
identical bottle full of water in the other hand. Hold each
bottle away from your body.

3 **Observe** Describe what you feel. Is each bottle pulled
toward Earth with the same force?

_____

_____

_____

4 **Infer** Is the amount of gravity on the two bottles the
same? How could you tell?

_____

_____

_____

Structured Inquiry

How does distance affect the pull of a magnet on metal objects?

Form a Hypothesis

You know that some metal objects, such as paper clips, are attracted to magnets. What happens when you change the distance between a magnet and a pile of paper clips? How does this affect the magnet's pull on the paper clips? Write a hypothesis. "If you move a magnet closer to a pile of paper clips, then . . ."

_____

_____

Test Your Hypothesis

1 Gather a pile of paper clips on your desk. Stand up a ruler near the paper clips.

2 **Experiment** Hold a magnet as shown below. Slowly lower the magnet until it is only 1 cm above the pile.

© Macmillan/McGraw-Hill

3 **Measure** Move the magnet away from the pile. Remove the paper clips and count how many stuck to the magnet. Record this data in a table.

Step **3**	
Distance	**Number of Paper Clips**
1 cm	
2 cm	
3 cm	

4 Repeat steps 1–3, holding the magnet 2 cm and 3 cm away from the pile of paper clips. Record your data.

_____

_____

Draw Conclusions

5 **Use Numbers** At what distance did the magnet pick up the most paper clips?

_____

_____

6 **Interpret Data** Does a magnet's pull on objects get greater or smaller as the magnet moves away from the objects?

_____

_____

Name _____ Date _____

Guided Inquiry

Can magnetic force pass through an object?

Form a Hypothesis

Can magnetic force pass through different objects, such as wood, plastic, paper, or foil. Write a hypothesis.

_____

_____

Test Your Hypothesis

Design a plan to test your hypothesis. List the materials you will use. Write down the steps you plan to follow.

_____

_____

_____

Draw Conclusions

Did any of the objects block magnetic force? Did any of the objects make the magnetic force stronger or weaker? Share your results with your classmates.

_____

_____

**Be a
Scientist**

Open Inquiry

What other questions do you have about magnets? For
example, what common objects are attracted to magnets.
Design an experiment to find out.

Remember to follow the steps of the scientific process.

My question is: _____

_____

How I can test it: _____

_____

My results are: _____

_____

Name _____   Date _____

What is work?

Make a Prediction

How do you know when work is being done?

_____

_____

Materials
- book
- chair

Test Your Prediction

1 Make a table like the one shown here.
Perform each action listed in the table.

Actions	Is It Work?	Why or Why Not?
pick up a book		
think about a problem		
slide a chair		
press feet against floor		
push against wall		

2 **Classify** Decide whether each action was work. Ask yourself if you got something done.

_____

_____

_____

_____

Draw Conclusions

3 Communicate Explain why you classified each action the way you did. Record this information in the table.

4 Infer What do you think work is?

_____

_____

Explore More

Experiment Perform other actions at home. Try to find actions where you do different amounts of work.

_____

_____

Open Inquiry

Is work done during sports? Think of a question about the activities done during a sport. Make a plan and carry out an experiment to answer the question.

My question is: _____

_____

How I can test it: _____

_____

My results are: _____

_____

_____

_____

© Macmillan/McGraw-Hill

Name _____ Date _____

What work is done while cooking?

Materials
- paper
- pencil

Purpose
Compare activities and decide which ones involve doing work and which ones do not.

Procedure

1 The list below shows several actions that you might perform when you bake cookies.
  a. Turn on the oven.
  b. Scoop flour out of a canister.
  c. Drop cookie batter on a baking sheet.
  d. Eat a cookie.

2 **Classify** During which of these actions is work done?

_____

3 **Explain** How do you know work is done during any of these actions?

_____

4 Make a list of at least six actions you did while getting ready for school this morning. Label each action as "work" or "not work." Include at least three actions during which work is done. Include at least three other actions during which work is not done.

_____

_____

_____

_____

© Macmillan/McGraw-Hill

Using Energy

1 You get energy to move and play from the foods you eat. Food is a source of stored energy. The table below shows how much stored energy is in some of the foods we eat.

Food	Calories of Energy
1 cup of apple juice	120
1 slice of wheat bread	75
1 slice of turkey	30
1 slice of cheese	60
1 cup of lettuce	7

2 **Use Numbers** Use the table to plan a meal. How many calories are in your meal?

_____

3 **Use Numbers** Choose an activity from the table below. How long can you do that activity before you have used up all the stored energy from your meal?

_____

_____

4 **Use Numbers** Choose another activity and repeat step 3. Which activity uses the most energy?

_____

Activity	Calories Used in 30 Minutes
biking (slow)	100
jogging	160
listening to music	17

Infer

When you do an experiment, you are trying to answer a question. Sometimes you can answer a question from the data you collect. Other times, you must **infer** the answer using facts you know.

Learn It

When you **infer**, you form an idea based on observations and facts. As you make observations, it is important to record your data. The more data you collect, the better you will be able to infer.

◀ A water wheel is a machine that uses the energy of moving water to power mills and factories.

© Macmillan/McGraw-Hill

Try It

Can running water do work? To answer this question, make a water wheel. Then observe what happens to it under running water. Use your observations and what you know about work to **infer** if water can do work.

1 Cut four 3–cm slits into a plastic plate. Then bend the slits to create a pinwheel.

2 Gently push a pencil through the center of the plastic plate. ⚠ **Be Careful!** Point the pencil away from your body. Ask an adult for help.

3 Tie one end of a piece of thread to a paper clip. Tape the other end to the pencil, near the hole in the plate.

4 Turn on the faucet so a little water flows out.

5 Rest the pencil across the palms of your hands. Then hold the edge of the plate 2 cm under the water. Record your observations.

_____

6 Repeat with a larger stream of water. Record what you observe.

_____

Now use observations and facts you know to answer these questions:

▶ What makes the water wheel move?

_____

Materials
• paper plate
• ruler
• pencil
• scissors
• thread
• paper clip
• tape
• faucet

▶ Does using more water give the wheel more energy? How can you tell?

_____

_____

▶ Can running water do work? Explain your answer.

_____

_____

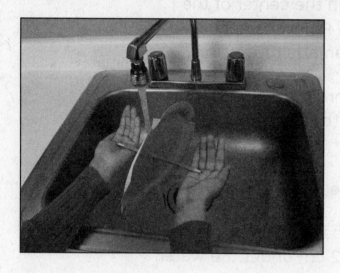

Apply It

You have learned to **infer** the answer to a question from the data you collect and the facts you know. Now you can infer answers to new questions. For example, can wind do work? How might you use your wheel to infer the answer?

_____

_____

How can a simple machine help you lift objects?

Form a Hypothesis

Look at the photo of step 2. Will moving the ruler's position on the marker change the amount of force needed to lift two blocks? Write a hypothesis.

_____

_____

_____

Materials
• clay 
• thick marker
• ruler
• 2 small cups
• large blocks
• one-gram cubes

Test Your Hypothesis

1. Use some clay to stick a marker to the center of a ruler. Then use clay to stick a small cup to each end of the ruler as shown in the photo.

Step 2

2. **Experiment** Put two large blocks in one cup. Add gram cubes to the other cup. How many cubes does it take to lift the two large blocks?

_____

3. **Use Variables** Change the position of the marker. Move it closer to one end of the ruler.

4. **Experiment** Repeat step 2. How does the marker's new position change your results?

Name _____ Date _____

Draw Conclusions

5 **Communicate** How does this machine lift objects?

_____

_____

6 **Interpret Data** How does the position of the marker change the number of gram cubes you need to lift the two large blocks?

_____

_____

Explore More

Experiment When are the two blocks lifted higher in the air—when the marker is near the two large blocks or when it is near the gram cubes? Try to find out.

_____

_____

Open Inquiry

How is the distance a force moves an object related to the amount of force? Think of your own question about this relationship. Make a plan and carry out an experiment to answer the question.

My question is: _____

_____

How I can test it: _____

_____

_____

My results are: _____

_____

© Macmillan/McGraw-Hill

How does a machine make work easier?

Materials
- book
- metric ruler

Purpose

Relate force and distance to the amount of work done.

Procedure

❶ Place a wooden metric ruler on the edge of a desk with half of the ruler hanging over the edge. For a standard metric ruler, about 15 centimeters will hang over the edge of the desk.

❷ Place a book on the end of the ruler that is on the desk.

❸ **Measure** Press down on the ruler until the book is raised 4 centimeters off the desk. Note how much force you had to exert.

_____

❹ Move the ruler until only about 10 centimeters of the ruler hang over the edge of the desk. Again, push down on the ruler until the book is raised 4 centimeters.

_____

❺ **Draw Conclusions** Look at your data and observations. Write a sentence that tells how work, distance, and force are related.

_____

_____

_____

Inclined Planes

1 Make an inclined plane as shown below. Then tie a bag of 25 marbles to a spring scale.

2 **Measure** Lift the spring scale straight up so that it is even with the height of the books. Record what the spring scale reads. Measure and record the distance you pulled the marbles.

3 **Measure** Use the spring scale to pull the marbles up the ramp. Record what the spring scale reads. Measure and record the distance you pulled the marbles.

4 **Interpret Data** Which method of moving the marbles required more force? Which method required moving the marbles a greater distance?

_____

_____

_____

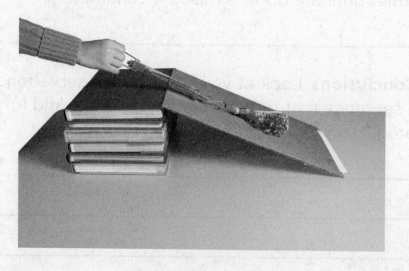

© Macmillan/McGraw-Hill

What happens to air when it is heated?

Form a Hypothesis

How does heat affect air? Does it make air get bigger or smaller? Write a hypothesis.

_____

_____

Test Your Hypothesis

1 Use a dropper to place five drops of water along the edge of a bottle's opening. Place a plastic disk on top of the opening. Then put the bottle in a refrigerator for several hours.

Step 1

2 **Predict** What will happen to the disk if the temperature of the air in the bottle increases?

_____

_____

3 **Observe** Remove the bottle from the refrigerator. Rub your hands together quickly. When your hands feel warm, place them on the bottle. Look at the disk.

Step 3

Draw Conclusions

4 **Communicate** What happened to the disk? Was your prediction correct?

_____

_____

5 **Infer** Think about what happened to the disk. What happens to air when it is heated?

_____

_____

Explore More

Experiment Place an empty plastic bottle in the refrigerator for several hours. Remove the bottle from the refrigerator and immediately stretch a balloon over the opening. What happens to the balloon?

_____

_____

Open Inquiry

What would happen to the balloon if the plastic bottle were placed in a sunny window? Think of your own question about what would happen to the balloon. Make a plan and carry out an experiment to answer your question.

My question is: _____

_____

How I can test it: _____

_____

My results are: _____

_____

Do your hands feel warm?

Materials

• 3 plastic bowls

• tap water

• ice

• thermometer

Make a Prediction

How do you think your hands will feel if you hold one in ice cold water, one in hot water, and then place both in warm water?

_____

_____

_____

Test Your Prediction

1 Fill one bowl half full with cold tap water. Add ice cubes to this bowl. Fill another bowl with lukewarm water from the tap. Fill the third bowl with hot tap water. ⚠ **Be Careful.** Do not make the water too hot. You do not want to burn yourself. Use the thermometer to measure the temperature of the water in each bowl. Record the temperatures below.

Bowl	Temperature (in °F)
Cold water	
Lukewarm water	
Hot water	

2 **Experiment** Place one hand in ice water and one hand in hot water and count to five. Remove both hands and place them in the bowl of lukewarm water. How do your hands feel?

_____

_____

Name _____ Date _____

Heating Water and Soil

1 **Predict** Which heats up faster, a cup of water or a cup of soil?

_____

_____

2 **Use Variables** Fill one cup with 150 mL of water. Fill another cup with 150 mL of soil.

3 **Measure** Put a thermometer in each cup and measure the temperature of the water and soil. Record the data.

4 **Experiment** Put the cups in a warm place. Record the temperature in each after 15 minutes.

5 **Use Numbers** Find the difference between the first and last readings of each thermometer.

_____

_____

_____

6 **Interpret Data** Which cup warmed up more? How do you know?

_____

_____

_____

Experiment

You just learned about heat. You read that an insulator is a material that does not allow heat to pass through it easily. How can you find out if something is an insulator? You can **experiment** to answer a question like this.

Learn It

When you **experiment**, you perform tests to answer a question. You make observations and collect data. Then you interpret the data to answer a question. When you experiment, it is important to test only one thing at a time. This helps you know what caused your results.

Try It

Experiment to find out which is the best insulator: paper, plastic, or foam.

1 Which material do you think will keep the ice cubes solid the longest: paper, plastic, or foam? Write a hypothesis.

_____

_____

_____

2 Put two ice cubes into each cup.

3 Cover each cup with plastic wrap. Use a rubber band to seal the wrap to the cup.

4 Place the cups in a warm place.

5 Observe the ice in the cups every ten minutes for one hour. Record the changes that you observe.

Now use your results to draw conclusions.

▶ In which cup did the ice cubes melt the slowest?

_____

▶ Which cup is the best insulator?

_____

Apply It

Now **experiment** to find out which is the best conductor of heat: aluminum, plastic, or wax paper. Remember that a conductor is a material that lets heat pass through it easily.

Repeat this experiment using three different types of wraps and three paper cups. Wrap aluminum foil around one cup, plastic wrap around the second cup, and wax paper around the third cup. Remember to record your observations.

_____

_____

_____

_____

Name _____ Date _____

How can you make sounds?

Materials

- goggles
- paper
- plastic ruler
- rubber band
- cardboard box

Make a Prediction

Look at the paper, ruler, and rubber band. What must you do to make a sound with each object?

_____

_____

_____

Test Your Prediction

⚠ **Be Careful!** Wear goggles.

1 **Observe** Hold a piece of paper by one corner. Wave it around. What happens?

_____

2 **Observe** Place a ruler on a desk. Extend half of it over the edge of the desk. Hold one end of the ruler down, and tap the other end. What happens?

_____

3 **Observe** Wrap a rubber band around a box. Pluck the rubber band. What happens?

_____

Draw Conclusions

4 What happened when you moved the paper, ruler, and rubber band?

_____

5 **Infer** Can you make a sound with the paper, ruler, or rubber band without making it move? Explain your answer.

_____

_____

6 **Infer** How are sounds made?

_____

_____

Explore More
Experiment Test ways to change the sound you made with each object. Try to make the sounds louder or softer, higher or lower. For example, try pulling the rubber band tighter and then plucking it. Record your results and the steps you follow.

_____

_____

Open Inquiry
How does the length of something change the sound it makes? Think of your own question about how things make sounds. Make a plan and carry out an experiment to answer your question.

My question is: _____

_____

How I can test it: _____

_____

My results are: _____

_____

Name _____ Date _____

Can a balloon help you hear?

Materials
• balloon

Make a Prediction

Do you think a balloon can help you hear? Write a prediction about what you think will happen if you listen through a balloon.

_____

Test Your Prediction

1 **Experiment** Work with a partner. Have your partner walk about 10 steps away from you. Then ask your partner to turn toward you and whisper something. Can you understand what your partner said?

_____

2 **Experiment** Blow up a balloon and tie the end closed. Now hold the balloon up to your ear. Ask your partner to whisper something again. Can you hear your partner?

_____

_____

Draw Conclusions

3 **Infer** Why do you think you could hear better through the balloon?

_____

_____

_____

_____

Changing Sounds

1 **Predict** How can you change the sound a straw makes?

_____

_____

2 Flatten one end of a straw. Then cut across the tip of this end as shown.

3 **Experiment** Cover your teeth with your lips and then blow hard through the cut end of the straw. Describe what you hear. Now blow softer. How does the sound change?

_____

_____

4 **Experiment** Now try using straws of different lengths. Remember to cut the tip before you blow into the straw. Describe what you hear. How does the sound change?

_____

_____

_____

Structured Inquiry

How does sound move through different types of matter?

Materials

- 3 plastic bags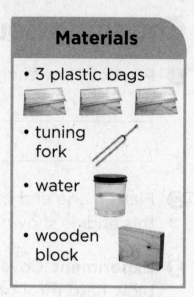
- tuning fork
- water
- wooden block

Form a Hypothesis

You just learned that sound travels through solids, liquids, and gases. How does the state of matter affect how sound travels? Write a hypothesis.

_____

_____

Test Your Hypothesis

1 Fill a plastic bag with air and seal it. Hold the bag against your ear.

2 **Experiment** Tap the tines of the tuning fork against the bottom of your shoe. Then hold the base of the tuning fork against the plastic bag. Listen to the sound it makes.

3 Fill a plastic bag with water. Seal it and hold it against your ear.

4 **Experiment** Tap the tuning fork and hold it against the bag. Record any differences you hear.

_____

5 Place a wooden block in a plastic bag. Squeeze out as much air as you can and seal the bag. Hold the bag against your ear.

© Macmillan/McGraw-Hill

6 **Experiment** Tap the tuning fork and hold it against the bag. How is the sound different now? Record your observations.

_____

_____

Draw Conclusions

7 How did the tuning fork sound different through the different materials?

_____

_____

8 **Interpret Data** Through which material was the sound loudest?

_____

_____

9 **Infer** Does sound travel best through a solid, a liquid, or a gas?

_____

_____

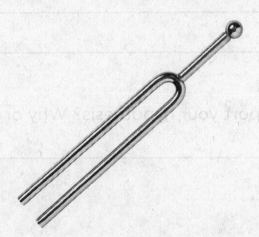

Guided Inquiry

How does sound move through different solids?

Form a Hypothesis

Sound can be stopped, slowed down, or absorbed, by different solids. How does sound travel through different solids?

_____

_____

_____

Test Your Hypothesis

Design an experiment to investigate how sound travels through different solids. Decide on the materials you will need. You may want to try plastic, wooden, and metal objects. Write out the steps you will follow. Record your results and observations.

_____

_____

_____

_____

Draw Conclusions

Did your results support your hypothesis? Why or why not?

_____

© Macmillan/McGraw-Hill

Open Inquiry

What other questions do you have about sound?
For example, what objects block sound the best?
Design an experiment to find out.

My question is: _____

_____

How I can test it: _____

_____

_____

My results are: _____

_____

Name _____ Date _____

How does light move?

Make a Prediction

What happens to light when it hits a mirror?

_____

_____

Materials
• mirror
• flashlight

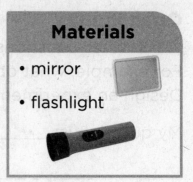

Test Your Prediction

1 Hold a mirror in front of you. Have a partner shine the flashlight onto the mirror.

2 **Observe** What happens to the flashlight's beam?

_____

_____

3 **Experiment** Pick a spot on the wall. Can you make light bounce off the mirror and shine on that spot? How? Do you have to move the mirror, the flashlight, or both?

_____

_____

Draw Conclusions

4 What happened to the beam of light when it hit the mirror? What happened when you moved the mirror? What happened when you moved the flashlight?

_____

_____

_____

5 **Communicate** Make a drawing to show how light moves when it strikes the mirror.

_____

_____

_____

Explore More

Experiment Sit next to your partner. Leave a meter of space between you and your partner. Then hold a mirror so that you can see your partner. Can your partner see you in the mirror? Can you see yourself and your partner in the mirror at the same time?

_____

_____

Open Inquiry

What would happen to the reflected light if the mirror were curved? Think of your own question about how light is reflected off a mirror. Make a plan and carry out an experiment to answer your question.

My question is: _____

_____

How I can test it: _____

_____

My results are: _____

_____

Name _____ Date _____

How do you see yourself in a spoon?

Make a Prediction

Do you think you will look different if you look at yourself in the two different sides of a metal spoon? Make a prediction about what you will see.

_____

_____

Test Your Prediction

1 **Observe** Look at yourself in the outside surface of a metal spoon, the side that cannot hold liquid. Describe what you see.

_____

_____

2 **Observe** Now turn the spoon over. Look at yourself in the inside surface of the spoon. Describe what you see.

_____

_____

3 **Infer** Why do you think your reflection looked different in different sides of the spoon?

_____

_____

© Macmillan/McGraw-Hill

Mixing Colors

1 **Predict** Look at the photo below and the color version in your textbook. What happens to the color of the plate when you spin it?

_____

_____

_____

2 Divide a white paper plate into eight equal parts. Color each section of the plate a different color.

3 **Observe** Carefully push a pencil into the center of the plate. Hold the plate away from your body. Spin it. What color do you see when the plate is spinning?

_____

_____

© Macmillan/McGraw-Hill

Name _____   Date _____

What makes a bulb light?

Materials

Make a Prediction

How can you connect a battery, a wire, and a light bulb to make the bulb light up?

- D-cell battery

_____

_____

- one 20-cm piece of insulated wire

_____

- light bulb

Test Your Prediction

1 **Experiment** Try to light the bulb using a light bulb, wire, and battery.

Step **1**

2 **Communicate** Draw each setup on a separate piece of paper. Record your results.

3 **Communicate** When your light bulb is lit, compare setups with your classmates. Is there more than one setup that lights the bulb?

_____

_____

_____

Draw Conclusions

4 How many setups could you find that made the bulb light?

_____

_____

5 **Infer** Look at the setups that lit the bulb. What do you think is necessary to make the bulb light up?

_____

_____

Explore More

Experiment How could you light two bulbs using only one battery? Can you think of more than one way? Try it.

_____

_____

_____

Open Inquiry

What would happen if one of the light bulbs in a circuit were broken? Think of your own question about how electric current flows. Make a plan and carry out an experiment to answer your question.

My question is: _____

_____

How I can test it: _____

_____

My results are: _____

_____

What does a balloon attract?

Make a Prediction

When you rub a balloon with a wool cloth, negative electrical charges are transferred from the cloth to the balloon. What kind of object do you think the balloon will attract?

_____

_____

Materials

- balloon
- wool cloth
- foam peanuts
- puffed rice cereal
- salt and pepper

Test Your Prediction

1 **Experiment** Blow up a balloon and tie the end closed. Rub the balloon for a few seconds with the wool cloth.

2 **Experiment** Hold the balloon near some foam peanuts. Are they attracted to the balloon? Then hold the balloon close to other materials. Are they attracted to the balloon? Record your results in the table below.

Materials Attracted to the Balloon	Materials Not Attracted to the Balloon

3 Was your prediction correct?

_____

_____

© Macmillan/McGraw-Hill

Conductors and Insulators

1 Put a battery in a battery holder. Connect a wire to each side of the battery holder.

2 Connect the free end of one of the wires to a socket that has a bulb in it. Then use a third wire and connect it to the socket as shown.

3 **Experiment** Gather objects, such as crayons and paper clips. Touch the free ends of the wires to each object.

4 **Observe** Does the bulb light up with each object? Record what happens.

_____

_____

_____

5 **Infer** Which objects are conductors? Which are insulators?

_____

_____

_____

Name _____ Date _____

Seasons

The change of seasons can be seen in many trees. In the winter, many trees have no leaves and are dormant. They are resting and do not grow during this time. In the spring, the trees produce leaves, flowers, and fruits. In the summer, the trees make and store food. In the fall, the trees start preparing for the winter and lose their fruits and leaves. Inside the branches and trunk of a tree, during the spring and summer, the tree produces more woody tissue and grows. From year to year, this growth is shown by a new ring of wood.

Materials

• small branch or tree

Purpose

Your task is to tell the age of a tree from its branch.

Make a Prediction

Look at the rings in a tree branch. Can you predict the age of a tree by adding up the number of rings?

_____

_____

Step 1

© Macmillan/McGraw-Hill

Test Your Prediction

1 Find a tree branch and count the number of rings.

2 Compare this branch with a thicker or thinner branch from another kind of tree. Does the new branch have more or fewer rings?

_____

_____

3 Compare the thicknesses of the rings in one tree branch. Do they vary in thickness? What might this tell you about tree growth?

_____

_____

Draw Conclusions

4 What did you observe? How old do you think the first tree you looked at is?

_____

_____

5 Is the second tree you looked at older or younger than the first tree? Why?

_____

_____

Critical Thinking

6 Do all places on Earth have seasons?

_____

_____

_____

7 What can annual rings in a tree tell you about the environment?

_____

_____

_____

Records from the Past

Fossils are the remains (leftovers) or traces (like a footprint) of a dead organism. A paleontologist is a scientist who studies fossils to learn about how life was on Earth in the past. A paleontologist goes out in the field and looks for fossils. When a fossil is found, the paleontologist carefully digs out the fossil with special tools. Then the paleontologist carefully transports the fossil back to the lab. At the lab, the paleontologist cleans the fossil and then studies the fossil to answer questions. What kind of animal did the fossil belong to? When was the animal alive? How old was the animal?

Materials

- plaster
- sand
- plastic animals
- bucket of water
- aluminum pan
- tools

Purpose

This activity will show how animal remains are buried and how paleontologists dig them out.

Form a Hypothesis

If animal remains are preserved when buried, write a hypothesis predicting how they might be studied.

_____

_____

Test Your Hypothesis

1 Mix sand, plaster, and water in a bucket or bowl. Use the following ratio: 3 parts sand, 1 part plaster, 1 part water. If the mixture is too dry, add a little more water, but make sure it is not too wet.

2 Pour half of the plaster/sand mixture into the aluminum pan.

3 Add various "fossils" on top of the plaster/sand mixture. Make sure that you spread them out.

4 Pour the remaining plaster/sand mixture on top of the fossils.

5 Let the fossil dig dry.

6 When the fossil dig is dry, you can start digging the fossils out with your tools. Carefully scrape away the plaster and use the brush to brush away the "dirt."

© Macmillan/McGraw-Hill

Draw Conclusions

7 What did you observe?

_____

_____

_____

8 What can you conclude about how paleontologists find fossils?

_____

_____

_____

Critical Thinking

9 Why do you only find fossil bones and teeth and not other parts of animals?

_____

_____

_____

10 In what way are your buried animals not like real fossils?

_____

_____

_____

The Water Planet

Seventy-five percent of the Earth is covered with water. Transportation on water is very common and important. Boats used to be built out of wood because wood floats. However, today, boats are made out of steel. A solid block of steel is too heavy to float. But if it is given a large, hollow, bowl-like shape, it can float.

Materials

• clay

• bowl of water

Purpose

Your task is to design and build a clay boat that will float on water.

Form a Hypothesis

How would you design a clay boat to float on water? State your hypothesis in the form of an "if, then" statement. For example, if the shape of the boat is . . ., then it will . . .

_____

_____

Test Your Hypothesis

❶ Take a small piece of clay and roll it into a ball.

❷ Shape another small piece of clay, that is about the same size, into a boat. Make sure the sides are high enough so that the boat does not flood with water.

❸ Put both pieces of clay in a bowl of water.

❹ **Observe** Observe what happens. If the boat sinks, reshape it until it floats.

Draw Conclusions

5 What did you observe?

_____

_____

_____

_____

_____

6 Did your observations support your hypothesis?

_____

_____

_____

_____

Critical Thinking

7 Why do you sink in a pool when you curl yourself up in a ball, but float when you stretch yourself out flat like a piece of paper?

_____

_____

_____

_____

_____

8 What would happen if more weight is added to a ship? Would it sink?

_____

_____

_____

Matter

Matter can either be a gas, a liquid, or a solid. Molecules make up matter. When many molecules are linked together in a special way, a polymer is made. The longer and more complicated a polymer, the harder it is for the molecules to move and the more likely it is that the polymer is a solid. Borax is a cross-linker. This means that borax links different chains of polymers together to make a more complex polymer.

Purpose

This activity will demonstrate how a new substance, such as a complex polymer, can be created from other materials.

Form a Hypothesis

When mixed, liquid glue and water form a polymer. If a cross-linker like borax is added, a more complex polymer should form. Write a hypothesis that states these facts in the form of an "if, then" sentence.

_____

_____

Test Your Hypothesis

1 In a plastic cup, mix $\frac{1}{4}$ cup of glue with 3 drops of food coloring.

2 Add 2 tablespoons of the borax solution (to make this solution, add 1 tablespoon of borax to $\frac{1}{2}$ cup of water).

3 Using the popsicle stick, mix all this together for about 30 seconds. This forms your slime.

4 Remove the slime and knead it with your hands for about 5 minutes.

5 To make a bouncy ball, squeeze as much water out as you can while kneading it. Then roll the slime between your hands to make a ball. Now watch it bounce!

Step **2**

© Macmillan/McGraw-Hill

Draw Conclusions

6 What did you observe?

_____

_____

7 Based on your results, what is your conclusion?

_____

_____

Critical Thinking

8 What do you think super-absorbent molecules are?

_____

_____

9 In Greek, *poly* means "many" and *meros* means "parts." What does *polymers* mean? Is this an appropriate name?

_____

_____

_____

Physical and Chemical Changes

Materials
- ice cube
- salt

Matter can undergo physical and chemical changes. After a physical change, the matter looks different but still has the same properties. After a chemical change, the matter has different properties and a new kind of matter is created.

Purpose

You will learn to tell the difference between physical and chemical changes.

Form a Hypothesis

When salt is sprinkled on ice, ice will melt faster. Write a hypothesis to state whether this is a physical change or a chemical change.

_____

_____

Step 1

Test Your Hypothesis

1 Sprinkle salt on an ice cube.

Draw Conclusions

2 What did you observe?

_____

_____

_____

3 Based on your results, what is your conclusion?

_____

_____

_____

Critical Thinking

4 Is the melting of ice with salt a physical change or a chemical change?

_____

_____

5 Can you explain why salt is used on the roads in areas where there is a lot of snow and ice during the winter?

_____

_____

_____

Energy

Although there are many different types of energy, all types of energy are used to perform some kind of work. The energy in food is called chemical energy. The energy used to turn on light bulbs is called electrical energy. Food and electricity contain potential energy (stored energy). When this potential energy is converted to kinetic energy (energy of motion), work is performed (such as using chemical energy from food to move muscles, using electrical energy to turn on a light bulb, and using energy from gasoline to drive a car).

Purpose

You will learn how potential energy can be turned into kinetic energy.

Form a Hypothesis

When a compressed spring is released, it expands and jumps around. Use this fact to form a hypothesis about the relationship between potential and kinetic energy.

_____

_____

© Macmillan/McGraw-Hill

Name _____ Date _____

Test Your Hypothesis

1 Place one end of a spring on the floor and press down on it.

2 Quickly take your hand away from the spring.

Draw Conclusions

3 What did you observe?

_____

_____

_____

4 Based on your results, what is your conclusion?

_____

_____

_____

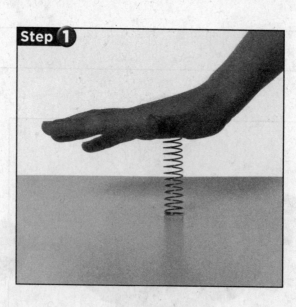

Step **1**

Critical Thinking

5 Do you think the kinetic energy of the spring once it is released is more or less than the potential energy before it was released?

_____

_____

_____

6 Give other examples of how energy is transformed from one kind of energy into another kind of energy.

_____

_____

_____

_____

_____

_____

Light

White light is made up of different colors of light. When white light hits an object like a prism, the colored light waves spread out and that is why we see a rainbow of colors coming out of the prism. White light can also be created by combining these rainbow colors.

Materials
• pencil
• white cardboard
• coloring pencils
• markers

Purpose

This experiment will prove whether white light is made up of the other colors of the spectrum.

Form a Hypothesis

The prism experiment shows that white light can be broken up into different colors of the rainbow. Write a hypothesis that uses a color wheel to test the reverse idea that the colors of the rainbow can be combined to create white light.

_____

_____

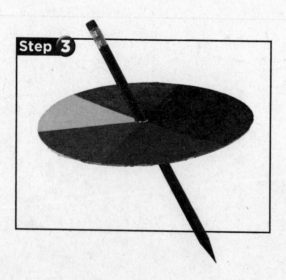

Test Your Hypothesis

1 Cut a circular disc with a diameter of 6 inches out of white cardboard.

2 Divide the white part of the cardboard into 6 equal pieces and color the parts as shown in the picture.

3 Push a sharp pencil half way through the center of the circle.

4 Hold the pencil with the point on the table and spin it fast.

Step **4**

Draw Conclusions

5 What did you observe?

_____

_____

_____

_____

6 Based on your results, what is your conclusion?

_____

_____

_____

Critical Thinking

7 How is a rainbow formed?

_____

_____

_____

8 What is refraction?

_____

_____

_____

© Macmillan/McGraw-Hill

Nocturnal Animals

Nocturnal animals are animals that are active at night. Many of these animals can see well in the dark. Humans, however, can not see well in the dark and that is why we use artificial light such as flashlights, car lights, and lamps to see at night. You probably have experienced that when you enter a very dark room it takes a while for your eyes to get used to the dark. In fact, it takes about half an hour for your eyes to fully adjust to the dark. If you used a regular flashlight to look at a star chart while you were stargazing, your eyes would need to adjust to the dark every single time you turned on the flashlight. The white light from the flashlight breaks down chemicals in your eyes that help you see in the dark.

Materials

- flashlight
- red cellophane
- rubber band
- star chart

Purpose

You will learn if using a flashlight covered with red cellophane will improve your vision at night.

Form a Hypothesis

If red light does not break down chemicals in the eye that help you to see in the dark, your eyes should adjust to the low light levels more quickly. Write a hypothesis expressing this idea.

_____

_____

Test Your Hypothesis

1 Turn off all the lights in your classroom, pull the window shades down, and look around with an ordinary flashlight, one that shines a beam of white light.

2 Turn the flashlight off and look around again. Is it easy to see objects?

3 Look around your classroom with a flashlight covered with red cellophane. Look around for at least half an hour.

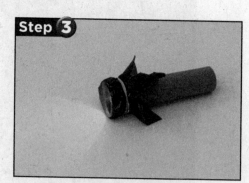
Step **3**

4 Turn the flashlight off. Can you see objects in the dark better than before?

Draw Conclusions

5 What did you observe?

_____

_____

6 Infer Why do you think it was easier to see in the dark after exposure to red light instead of white light?

_____

_____

_____

Critical Thinking

7 Are all stars yellow?

_____

_____

_____

_____

8 What is a shooting star?

_____

_____

_____

_____

© Macmillan/McGraw-Hill

Structured Inquiry

How does color help living things survive?

Materials

- wrapping paper
- 2 black and white pages from a newspaper
- scissors

Ask Questions

Flowers need to be noticed. They need animals to find them and help them reproduce and spread seeds. Does the color of a flower help the flower advertise itself?

Make a Prediction

What colors help a flower get noticed? Do certain colors work better than others at attracting animals in different places?

_____

_____

Test Your Prediction

1. Decide on a typical shape for a flower blossom. The overall size should be about 2 cm or less in diameter. Draw 10 flowers on a wrapping paper page and another 10 flowers on a black and white page. Cut them out.

2. While your back is turned, have a friend spread out the 20 cut-out flowers on a black and white page.

3. **Experiment** When your friend says "go," turn around and pick up as many flowers as possible in just 3 seconds. Pick up one flower and place it on the table before you pick up another flower. Do the test 3 times.

4 **Record Data** Record the number of flowers you picked up that were colored and the number that were black and white.

Test	Number of Colored Flowers	Number of Black and White Flowers
1		
2		
3		

5 Trade roles with your partner and repeat steps 2–3. Record your results.

Test	Number of Colored Flowers	Number of Black and White Flowers
1		
2		
3		

6 **Record Data** Use your data to make a double bar graph showing how many flowers of each color were picked up each time you did the test.

7 Repeat steps 1–6, and this time, place the flowers on a colored page.

My Results

Test	Number of Colored Flowers	Number of Black and White Flowers
1		
2		
3		

My Partner's Results

Test	Number of Colored Flowers	Number of Black and White Flowers
1		
2		
3		

Graph:

© Macmillan/McGraw-Hill

Communicate Your Results

Have a class discussion and share your results and graphs. What did you find out? Use your data to answer these questions:

▶ Which flowers were easiest to spot? Did the color of the sheet of paper make a difference?

_____

_____

_____

▶ Which flowers were hardest to spot? Did the color of the sheet of paper make a difference?

_____

_____

_____

▶ If flowers grew in a place where everything was brightly colored, what color flowers would be most easily noticed?

_____

_____

_____

▶ How do colors help flowers get noticed? What evidence supports your idea?

_____

_____

© Macmillan/McGraw-Hill

Name _____ Date _____

Guided Inquiry

Hiding in Plain Sight

Materials

- crayons, markers, or colored pencils
- scissors
- clear tape

Ask Questions

Hiding helps some animals avoid being eaten. Some animals need to hide so that the animals they want to eat do not see them waiting for a meal. How are some animals able to hide but still be right in front of us?

Make a Prediction

If hiding from other animals helps living things survive, make a prediction about how animals can use color to survive.

_____

_____

Test Your Prediction

1. Draw a simple picture of a lizard or a frog as if you were looking straight down on it from above. Color your drawing carefully so that it can hide in the classroom in plain sight. The drawing should be 10 to 15 cm long.

2. When everyone's drawings are colored and cut out, you and half of your classmates should tape drawings on the surfaces you have chosen. Do not hide them under or behind anything. The other half of your classmates should go out of the room while you are hiding the drawings so they cannot see where you are putting them.

Step 2

3 After all of the drawings are hidden, your classmates will try to find and list as many of the drawings as possible in 1 minute.

Communicate Your Results

Work in groups of 4 to 8 and discuss what you found out about hiding in plain sight.

4 How many pictures were hidden? How many pictures did each student find?

_____

_____

5 Were some pictures easier to find than others? Describe the color of the drawing and the color of the background of 1 picture that was easy to find. Describe the color and background of a picture that was hard to find.

_____

_____

_____

6 How do colors help animals hide? What evidence supports your idea?

_____

_____

Open Inquiry

Now You See It, Now You Don't

Invent and test other ways to explore showing off or hiding. Design and perform an experiment. Ask a question, make a prediction, test your prediction, record your data, and communicate your findings. Make a poster to show what you did and what you found out. Here are some ideas to get you started:

▶ Take a showing-off and hiding survey of living things in your schoolyard. What living things in your schoolyard are most easily seen? What organisms are not easily seen? What makes them good advertisers or hiders?

▶ In addition to color, what structures help organisms advertise or hide? Does shape make a difference?

My question is: _____

_____

How I can test it: _____

_____

_____

My results are: _____

_____

Structured Inquiry

How can you track the Sun and the Moon?

Ask Questions

How do shadows made by sunlight change during the day? Do they change in length during the day? Do they point in different directions? How can we explain the changes?

Make a Prediction

How do shadows change in length and direction with the time of day? Write a prediction.

_____

_____

Materials

• 1 pencil or other long, thin stick

• 1 piece of clay large enough to hold the stick

• 1 piece of white construction paper

• 2 magic markers (1 blue and 1 red)

Test Your Prediction

1 Place the construction paper in a south-facing window, or outside where you can see the Sun all day, with the pencil upright in the clay in the middle of the paper.

2 Place the clay in the middle of the paper. Stand the pencil in the clay.

3 Mark the shadow of the pencil with the red marker by drawing a line on the paper.

4 **Predict** Where will the shadow be in 1, 2, 3, and 4 hours? Draw lines with a blue marker to show your predictions.

Step **3**

5 Make certain that no one moves the paper during the day. As the shadow moves, trace its position with the red marker. Compare the actual positions outlined in red with your predictions marked in blue. Repeat the experiment a few days later.

Communicate Your Results

6 How did the shadow change in length? How did it change in direction?

_____

_____

_____

_____

_____

7 Share your shadow recording with others and post them in the classroom.

8 Write a story about how your shadow changed in length and direction. Were your predictions better on the second day than the first? How close were your predictions?

_____

_____

_____

_____

9 Have a class meeting and try to explain what made the shadows change.

_____

_____

_____

_____

Guided Inquiry

Moon Tracking

Materials

• 12-inch ruler

Ask Questions

Does the Moon appear to move across the sky like the Sun does? How would you find out?

Make a Prediction

Earth's rotation causes the Sun to appear to rise, move across the sky, and set. Do all objects in the sky move in a similar way? Write a prediction about the Moon's movement.

_____

_____

Test Your Prediction

1. Go outside with a parent or teacher at a time when the Moon is visible in the sky. Good times to do this would be late afternoon or early evening.

2. **Observe** Observe the Moon. Hold one arm straight out toward the horizon. Hold the other arm straight and point it toward the Moon. Draw the angle between your arms on a piece of paper. Have your parent or teacher help you. Record this measurement on a separate piece of paper.

Step 2

© Macmillan/McGraw-Hill

3 Repeat step 2 one and two hours later.

Communicate Your Results

4 Did the Moon's position change over time? Did the Moon rise or fall in the sky?

_____

_____

5 During the time you observed it, did the Moon change its phase?

_____

6 Do your results support your prediction? Can you explain the motion, if any, of the Moon across the sky?

_____

_____

Name _____ Date _____

Open Inquiry

More Moon Observations

Be a Moon watcher. How else can you track Moon changes?
Ask a question, make a prediction, set up an investigation,
record your data, and communicate your findings. Here are
some suggestions to get you started:

▶ Can you explain what causes the different phases of
  the Moon by making a model of Earth, the Moon, and
  the Sun?

  _____

  _____

▶ At night, does the moonlight cast a shadow? Do
  shadows from the moonlight change like shadows from
  the Sun?

  _____

  _____

  _____

▶ Earth's gravity holds the Moon in orbit around Earth.
  What effect does the Moon's gravity have on the Earth?

  _____

  _____

▶ Can you explain why we only see one side of the Moon?

  _____

  _____

  _____

© Macmillan/McGraw-Hill

▶ Observe the Moon at night just after a new Moon. Ask your teacher when the Moon is a new Moon. Go outside with an adult and watch changes from night to night over 2 weeks. The Moon will be seen in the west above where the Sun sets. How did the Moon change in an hour? How did it change in 1 to 3 days if viewed at the same time each day?

My question is: _____

_____

How I can test it: _____

_____

_____

_____

_____

My results are: _____

_____

_____

_____

_____

Structured Inquiry

How can you measure light energy?

Materials

- 2 clean socks: 1 black and 1 white

- 2 clear plastic cups or 2 500–mL beakers

- 20 drops of blue food coloring

- 2 thermometers

Ask Questions

How can we collect solar energy from the Sun? How would we know if we actually collected it?

Make a Prediction

If objects of different colors absorb different amounts of light energy, it should be possible to measure how much energy they absorb by measuring how warm they become. Write a prediction about what you think will happen to differently colored objects exposed to sunlight.

_____

_____

© Macmillan/McGraw-Hill

Test Your Prediction

1 Place the black and white socks on your hands as if you were putting on gloves. Hold your hands in direct sunlight for 2 to 5 minutes. Predict what you think the socks will feel like when they are held in the Sun.

_____

2 Perform this test and record your observations. Does one sock feel warmer than the other sock?

_____

3 Fill 2 clear plastic cups with cold water. Add 15 drops of blue food coloring to one of the cups (until the water is dark blue). Put an aluminum foil cover on each cup. Make holes in the foil covers and push the thermometers into the water.

Step **3**

4 Record the temperature of the water you put in the cups.

5 Set both cups in direct sunlight. Place them on a piece of white cardboard or foam.

6 Measure the temperature of the water in each cup every 5 minutes for 1 hour. Make a line graph to show the temperature change in each cup.

7 Repeat your experiment and compare the results to the first experiment. Record your results on the graph below.

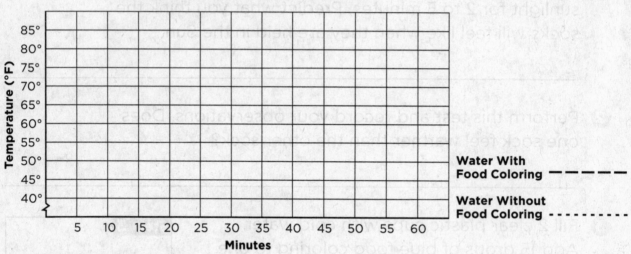

Temperatures of Colored Water

Water With Food Coloring — — — —

Water Without Food Coloring - - - - - - - - -

Communicate Your Results

8 What happened to the socks when they were placed in sunlight? What do your experiments suggest about how energy from the Sun was being collected and stored?

_____

_____

9 How do the cups of clear and colored water act like light and dark socks when placed in the Sun? What do you think is causing the difference in temperature in the samples of water?

_____

_____

_____

Guided Inquiry

Blocking Light

Ask Questions

How are shadows made? How can we change how shadows look? What are shadows made of?

Materials

- 1 flashlight

- 2 objects from around the classroom to use for making shadows

- paper on which to trace shadows that you make

Make a Prediction

To create a shadow, some opaque objects must block light that is shining toward them. Use this fact to make a prediction about what causes shadows and how their shapes might be changed.

_____

_____

_____

Test Your Prediction

1 Pick 2 objects of different shapes that you will use to make shadows.

2 Decide how you will make a shadow by using a flashlight and one of the objects. Record how far apart the flashlight, object, and wall will be from each other.

3 On a piece of paper, draw what you think the shadow of the object will look like and how big it will be when you use a flashlight to make a shadow from the object.

4 Now make a shadow with the flashlight, and draw the actual shadow on the piece of paper where you drew your prediction.

5 Try to make different shaped shadows with the same object. Draw your results. Record what you did to change the shape of the shadow.

6 Make a shadow on a wall using an object and the flashlight. Is there a shadow in the air between the object and the image on the wall? How can you tell?

_____

_____

_____

Communicate Your Results

7 Share your shadow drawings of your prediction and the actual results of making a shadow. Make a rule that would tell others how shadows are made and how we can change the shape or size of a shadow.

_____

_____

_____

8 Have a shadow art show and post drawings of actual shadows you made by moving an object into different positions. Have other students guess what object was used to make each shadow.

9 With classmates, discuss what shadows are made of. Are they made by light? The object? The wall? Is the shadow only on the wall or floor? What is the evidence for your answers?

_____

_____

_____

Open Inquiry

More Brilliant Experiments

What questions about collecting or blocking light do you have? What experiments would you like to do to find out more about light? Here are some ideas:

▶ What kinds of shadows can you make using 2 flashlights and one object?

▶ What is the biggest shadow you can make in the classroom with an object the size of a baseball?

▶ How could you build a new sunlight (solar energy) collector? Could you use a shoebox or a plastic trash bag? How about a coffee can?

Design an experiment based on your questions. What else do you want to explore? Ask a question, make your prediction, plan an investigation, perform the experiment, record your data, and communicate your findings. Make a poster to show what you did and what you found out. What did you observe?

My question is: _____

_____

How I can test it: _____

_____

My results are: _____

_____

_____

© Macmillan/McGraw-Hill